IN-N-OUT
BURGER

IN-N-OUT BURGER

A BEHIND-THE-COUNTER
LOOK AT THE FAST-FOOD CHAIN
THAT BREAKS ALL THE RULES

Stacy Perman

COLLINS BUSINESS
An Imprint of HarperCollinsPublishers

HarperCollins books may be purchased for educational, business, or sales promotional use. For information, please write: Special Markets Department, HarperCollins Publishers, 10 East 53rd Street, New York, NY 10022.

Designed by Kris Tobiassen

Library of Congress Cataloging-in-Publication Data
    Perman, Stacy.
        In-N-Out Burger / by Stacy Perman.—1st ed.
            p.   cm.
        ISBN 978-0-06-134671-2
        1. In-N-Out Burger (Firm)—History. 2. Fast-food restaurants—United States—History. I. Title.
TX945.5.I5P47 2009
647.9573—dc22

                                                                2008035605

09 10 11 12 13   OV/RRD   10 9 8 7 6

"The destiny of nations depends on the manner in which they nourish themselves."

—Jean Anthelme Brillat-Savarin,
The Physiology of Taste (1854)

"Keep it real simple. Do one thing and do it the best you can."

—Harry Snyder, co-founder
of In-N-Out Burger

PROLOGUE

On the dry, desert morning of April 24, 2007, the sky swept clean of clouds, In-N-Out Burger opened its 207th restaurant in Tucson. Located at the edge of the El Con Mall on East Broadway, not far from the campus of the University of Arizona, the opening was the fabled California chain's fourteenth entry in the state. Almost immediately a boisterous crowd appeared, requiring the presence of police officers to direct traffic and help with crowd control. While a swirl of excitement usually accompanied new store openings, the Tucson kickoff seemed to generate an unprecedented level of hysteria. For the first time in years, the area around the depressed and largely vacant El Con was bustling. The pent-up demand for In-N-Out marked a drastic change for the city's oldest shopping center, which had been languishing for so long that a local newspaper had once described its "deserted core" as filled with little more than an "eerie stillness."

As customers descended upon In-N-Out's freshly paved drive-through, the scene quickly took on the air of a noisy parade. There were businessmen in suits, women in heels, truckers in jeans, college students in T-shirts and with pierced noses, construction workers in heavy boots, and moms with babies on their hips. They all braved the chain's infamously long lines, enduring waits of more than two and three hours. In the sky overhead, news helicopters whirled, capturing the clamor in the parking lot below. After witnessing the ensuing frenzy from the ground, a local science fiction author named Matt Dinniman was later moved to remark: "If you actually drove by the place today, you'd think Jesus himself was working the shake machine."

Remarkably, the official kickoff was marked with little fanfare. Two workers carried a simple white sign down East Broadway with the familiar In-N-Out logo—a yellow boomerang arrow—and written in block lettering the words "WE ARE OPEN," and propped it up on a pair of wooden sawhorse legs. The associates fired up the grill, unlocked the doors, and opened for business. There weren't any strings of colored plastic flags fluttering in the wind to herald the arrival of Tucson's very first In-N-Out; there weren't any prizes offered, furry mascots, or any of the other marketing gimmicks that usually lure customers to new fast-food openings. There was no promotional advertising either—just a small sign that stood on the lot for some time. "Coming Soon, In-N-Out Burger."

In fact, there was no "grand opening," at least not in the traditional sense; the carnival atmosphere was created entirely by In-N-Out's rabid fans. The chain didn't need to advertise its opening; for weeks, its devotees had been broadcasting the news to one another.

For two years, ever since word had spread that an In-N-Out was coming to their city, residents had been waiting anxiously. As April 24 approached, a frantic excitement overtook Tucson and the surrounding suburbs. Despite months of speculation and press inquiries, the company, while friendly, revealed little—so a number of Tucsonans began their own campaign to uncover any and all signs of progress. Driving by the site regularly, they documented sightings of a coat of fresh paint or the pouring of concrete and broadly circulated their findings. When the day finally came that the Tucson In-N-Out was no longer just a matter of hope and rumor, it felt like Christmas in April. "There are so many people excited about these stores coming to town," was how Dave Smith, a local real estate broker, described the thrill that blanketed the city. "It is like they are almost willing them out of the ground."

For years, Arizona residents regularly drove across state lines in order to eat one of In-N-Out's vaunted Double-Double burgers.

When In-N-Out opened its first Arizona store in the city of Scotts-dale in 2000, Tucsonans got in their cars and made the ninety-mile journey. While Scottsdale might not be considered convenient, it was certainly closer to Tucson than the Lake Havasu City location that opened the same year (which at a distance of 321 miles from Tucson was an unreasonable distance for a burger run—although not an en-tirely uncommon one). In fact, until In-N-Out first arrived in Arizo-na, many residents flew roundtrip between Phoenix and the Ontario International airport in California expressly for a $2.75 hamburger.

The Tucson debut was handled not unlike a presidential visit or a movie premiere. Prior to the official opening, local dignitaries and members of the city council joined In-N-Out executives at an invitation-only, pre-opening party that also served as a final run-through for the associates. Tucson's media covered the official opening; the *Arizona Daily Star* sent a team of six journalists on an assignment that they dubbed "Operation In-N-Out."

Ravenous customers began arriving in the dark of night, long be-fore the store's 10:30 a.m. opening. Actually, people began lining up at 2:00 a.m. the day before, some sleeping in their cars overnight. By 6:00 a.m. on the morning of the opening, about a dozen folks were pressed against the front doors. By noon, the crowd had grown ap-preciably, hundreds of people having descended upon the fast-food restaurant and its signature crossed palm trees. Marveling at the thick, snaking procession of people, Phil Villarreal, a reporter for the *Arizona Daily Star*, recalled soviet-era bread lines in Moscow.

As the day wore on, the walk-in line continued to expand like an unencumbered waistband. By midday it had grown so dense, at times one hundred customers strong and six and seven people deep, that the entire line had taken over a neighboring parking lot and yel-low plastic police tape was used to rope off the crowd in as orderly a fashion as possible. As the mercury rose, bright-eyed and smiling In-N-Out associates passed out cups of water and pink lemonade to help relieve both the tedium and the heat for the customers waiting in line. After inspecting what could only be described as a stampede, In-N-Out's vice president of planning and development Carl Van

Fleet assessed the scene with the chain's typical understatement, telling the *Daily Star*, "This is not something that happened overnight. I think it just sort of grew with us." As night fell, the crowd showed little sign of thinning out. At nearly 10:00 p.m., there were still over one hundred cars in the drive-through lane.

The company temporarily opened a mobile kitchen on-site to help with the constant demand. In-N-Out had already sent in a team of about forty veteran employees (called the "In-N-Out All-Stars") from existing locations to help the two dozen new associates work the opening. Like a mobile commando unit, the All-Stars are highly skilled and experienced In-N-Out veterans, trained to be able to handle any situation under pressure. Dispatched to every new store opening since about 1988, their job was and is to guarantee a smooth debut. But even with its own precision planning, military execution, and its All-Stars on hand, In-N-Out Burger could have had no idea of the kind of frenzy that the Tucson opening would create.

Drivers sat in their cars for hours, waiting to place their orders. The drive-through line stacked up for blocks, ran out onto the street, and overflowed into the El Con Mall's parking lot, snarling up traffic. A nimble associate wearing a menu affixed to a chain went from car to car taking orders and sending them back to the kitchen via PDA in an attempt to help speed things up.

Others abandoned the pile-up of cars altogether and made the pilgrimage by foot. A swell of people had already formed inside the restaurant; the overflow spilled outside, long past the glass entry doors into controlled chaos. A respectful rush of customers inched their way toward the gleaming stainless steel counter where a group of smiling workers in their crisp white uniforms, red aprons affixed with large silver kilt pins, and paper cadet hats awaited them. Remarkably, despite the constant crush of customers, the associates maintained their smiles throughout the grueling day.

Under a yellow neon sign that spelled out in cursive letters the company's motto—"Quality You Can Taste"—those at the front of the line ordered from In-N-Out's famously limited menu of three burger items, french fries, soft drinks, lemonade, and milkshakes—a menu that has barely changed since Harry Truman was president. There were no Mediterranean wraps, Caesar chicken salads, or children's menus. Facing the antiseptically clean open kitchen, customers saw that there were no heat lamps, freezers, or microwaves, and the heavy odor of grease and meat was curiously absent. There were no bags of flash-frozen fries on-site either. Rather, in a procedure that has gone unchanged since the chain first opened in 1948, a cheery associate hand-peeled, cut, and fried the raw Kennebec potatoes grown especially for the chain. At In-N-Out, the lettuce is still leafed by hand, the sponge-dough buns are baked daily—and both the cheese and the ice cream for its shakes remain 100 percent dairy.

After braving the lines, some ordered forty or fifty burgers to bring back to friends and co-workers. One customer, Judi Esposito, arrived in hand with a stockpile of gift coupons worth forty dollars that she had collected for the past year and half in anticipation of the opening. "It sounds crazy," she told a local reporter, simply thrilled to finally order a meal from In-N-Out on her home turf, "but they're just really good, sloppy hamburgers."

Those in the know ordered off-menu from what has become known as In-N-Out's "secret menu." An insider code, it is an unofficial parallel menu that has traveled from Southern California to Northern California and then on to Nevada and Arizona and even beyond In-N-Out's traditional borders. Although nobody knows its exact origins, the secret menu has existed for decades, and knowledge of Animal Style, the 4x4, and the Flying Dutchman has been passed on primarily through word of mouth (although in recent years the secret menu can also be found on In-N-Out's website).

The Tucson opening was In-N-Out's busiest to date. In the days and weeks following the launch, the massive lines and pileups subsided only slightly. A month after the opening, fans reported that the drive-through lane dropped from one hundred cars to fifty, with the waiting time inside averaging about thirty minutes from order to pick up.

After the enthusiastic reception given to the Tucson store, when on the morning of November 15, 2007, In-N-Out Burger unveiled its second Tucson-area store in the suburb of Marana near Interstate 10, the local NBC affiliate was moved to report, "Big news for Marana today. The town gets its first In-N-Out Burger. . . . So not only are we getting bigger around the waistline, so is the area."

Although the scene was decidedly tamer than the one that consumed the Tucson opening seven months earlier, it was no less anticipated or exciting. Bracing themselves against the hungry crowds, the Marana police set up a command center near the shop on 8180 North Cortaro Road and five off-duty police officers (hired by In-N-Out) were assigned to direct traffic. Half a dozen people staked out places near the front doors in the early morning hours, hoping to be among the first in line.

Available only in California, parts of Nevada, Arizona, and more recently Utah, the opening of a new In-N-Out Burger has long been cause for outsized celebration. When an In-N-Out opened in suburban San Francisco in 2001, there were epic lines for months. In 2003, the anticipated Millbrae opening was delayed because city officials were concerned about the kind of traffic jams that In-N-Out would bring. When in 2007 *Departures* magazine printed the unconfirmed rumor that In-N-Out Burger was scouting locations in New York City, excited East Coast fans chased down the report, frantically dialing the company's toll-free telephone number.

Golfers and groupies on the PGA Tour cheered when the Scottsdale store opened. Not far from the TPC Scottsdale Golf Course (where the $6 million purse FBR Open is played), the pros began having In-N-Out burgers delivered directly to them following their rounds. When Jonathan Kaye won the tournament in 2004, his pro-am partner was in such a hurry to hit the In-N-Out that he forgot to turn in Kaye's scorecard and ended up calling it in from the drive-through lane. "It was crazy," Greg Wolf, the course's head professional, later exclaimed. "There were cops directing traffic there for the first month."

And when, on the eve of the Fiesta Bowl in 2007, a reporter asked Ohio State quarterback Troy Smith if the Buckeyes had any advantage over playing Florida State at Sun Devil Stadium (since it was the team's fourth trip to Phoenix in five years), the Heisman Trophy winner replied, "The first thing I am most familiar with is In-N-Out Burger, which is around the corner from our hotel. Probably the height of my day every day is getting a chance to go to In-N-Out Burger. I love those cheeseburgers."

The little regional chain that was built on the philosophy of quality, made-to-order hamburgers, and "the customer is always king" had over the years drawn fans from every imaginable quarter. In an industry that has come to be seen as a scourge on modern society, responsible for everything from obesity to urban blight to cultural imperialism, this modest, low-slung eatery with the big yellow arrow is unique among fast-food breeds: a chain revered by hamburger aficionados and epicureans, anti-globalization fanatics and corporate raiders, meat-eaters and even vegetarians. Make mention of the three monosyllabic words and a kind of reverie takes hold. People's eyes close and their lips begin to quiver with the pleasures of sense memory. "For years, In-N-Out was the other woman in my life," my mother once told me. "I'd be cooking dinner and your father would go to an In-N-Out on his way home from the office. He'd eat a Double-Double, fries, and a chocolate shake. I'd have dinner on the table, and he'd say 'I'm not hungry.'"

The family-owned, fiercely independent chain has remained virtually unchanged since its inception in 1948. It is the envy of the industry and the darling of investment bankers, who routinely put In-N-Out on their IPO wish list. In fact, by the end of Eric Schlosser's screed against the industry, *Fast Food Nation*, as nearly every other chain is pilloried and left in a heap, In-N-Out Burger remains standing virtually unscathed. It has long been adored by its legions of fans, whose main complaint seems to be that there just aren't enough of the restaurants around.

With little prodding, In-N-Out has earned the ringing endorsement of an astonishing spectrum of people. Rocker Courtney Love reportedly insisted on visiting an In-N-Out just before entering rehab in 2004. A year earlier, Julia Child—the grand dame of French cuisine—dispatched her assistant to the chain in order to sate her craving while she was in the hospital recovering from knee surgery.

During the 2004 NBA finals (the Los Angeles Lakers against the Detroit Pistons), California governor Arnold Schwarzenegger had a wager with Michigan governor Jennifer Granholm in In-N-Out burgers over the outcome of the game.

Singer Beyoncé Knowles admitted to making an In-N-Out run before the 2007 Academy Awards show. Following the ceremony, Oscar winner Helen Mirren was photographed tucking into an In-N-Out burger in her custom Christian Lacroix gown at *Vanity Fair*'s post-Oscar party. Since 2001, the magazine's editor, Graydon Carter, has rented one of In-N-Out's cookout trailers for the annual fete. "The stand has always been a huge hit," he explained. "Many of the people who come from the ceremony are ravenous and make straight for the stand. Plus, many of the women have been fasting for weeks to get into the dresses they are wearing."

Over the years, the relationship between Hollywood and In-N-Out Burger has only intensified, becoming nearly as much a part of celebrity culture as fan magazines, autographs, or publicity stills. When discussing his favorite places to dine in Los Angeles (for an interview promoting his film *Collateral*), Tom Cruise listed In-N-Out Burger alongside the longtime hot spot for the well-heeled, Spago. While filming *The Green Mile*, Tom Hanks rented an In-N-Out cookout trailer for the set. And sly In-N-Out Burger references have found their way into a number of television programs and films from the Coen brothers' cult hit *The Big Lebowski* to *The Simpsons*—all of which prompted the august *New York Times* to ask, "What's so hip about a hamburger chain?" on its cultural pages.

Displaced Californians pine for In-N-Out's Double-Double burgers; most of them can sing the chain's radio jingle. "In-N-Out, that's what a hamburger's all about," with greater confidence and accuracy than they can the national anthem. Travelers are known to time their itineraries to hit an In-N-Out during a mealtime, and its most ardent fans enjoy comparing notes on how far they have gone to find an In-N-Out. When British-born Kelly Grant worked as a nanny, she used to take her five- and seven-year-old charges to the In-N-Out Burger by Los Angeles International Airport. Then she would take them across the field where the trio would eat their burgers and plane spot. "It was our favorite thing to do," she explained.

In 1999, In-N-Out made international headlines after U.S. army sergeant and Baldwin Park native Andrew Ramirez told a phalanx of reporters that what he craved most after being held as a prisoner of war in Kosovo by Serbian forces for thirty-two days was a Double-Double. Ramirez's mother, Vivian, made headlines of her own when she flew to see him at his German army base following his release, carrying with her two of the burgers and an order of fries.

In-N-Out's fan-customers can usually remember the first time they ate at an In-N-Out much in the same way that most people can remember their first kiss. "I grew up in Detroit," explained Robert LePlae, the president of advertising agency TBWA/Chiat/Day. "I remember it was on my first trip to California in 1984 on a production shoot. A buddy of mine from college took me to the In-N-Out in Fullerton. A year later I moved to California, and it was the first meal I had off the plane. I thought about In-N-Out Burger all year."

In fact, when famed restaurant The French Laundry reached its twelfth anniversary in 2006, its Michelin-starred chef Thomas Keller, a longtime fan, celebrated with his staff by having three hundred In-N-Out burgers and a small mountain of fries delivered to the restaurant. That same year, in something of a tribute, Keller, was photographed in the April issue of *Food & Wine* magazine sitting at a booth at the Napa Valley In-N-Out wearing one of the chain's paper cadet hats, the restaurant's manager sitting across from him.

Frequently the subject of rumor and speculation, the fastidiously private company has always shunned the kind of publicity that its competitors routinely courted. And while it has rarely bothered to counter or clarify the murmurings, the conjecture has touched on everything from the recipe for its secret sauce to the meaning of the twin palm trees planted in a cross formation at each store—there has even been speculation as to whether the company was run by some kind of Christian cult. Every so often, word that In-N-Out Burger is about to franchise or be sold makes the rounds. While this kind of business gossip has been kicking around for decades, all its customers usually want to know is how soon an In-N-Out Burger will open near them.

During the sweltering summer of 2006, the rumor mill went into overdrive. In early August, In-N-Out Burger issued a statement announcing that Esther Snyder, who (along with her husband, Harry Snyder) had founded the chain, had died. She was eighty-six years old. Esther's death left her twenty-four-year-old granddaughter Lynsi Martinez the sole heir to the In-N-Out Burger fortune.

A petite, unassuming woman, Esther Snyder had spent most of her life determinedly maintaining In-N-Out as a private, family-owned business, keeping it virtually unchanged as decades rolled by. She did so despite countless offers from investment bankers, venture capitalists, private individuals, and large corporations, all of whom were hellbent on buying into the hugely profitable chain's unparalleled popularity. Esther Snyder preserved her family's privacy even as In-N-Out transformed from a single burger stand to a cult phenomenon.

Her death sparked a round of speculation about the future of the counterintuitive chain that had stridently bucked every industry trend. It was exactly the kind of speculation that had only recently begun to die down. Throughout part of 2005 and much of 2006, an ugly lawsuit played out in Los Angeles Superior Court in which it was alleged that Martinez and her brother-in-law Mark Taylor were plotting a coup to oust Esther from the company. For months, a series of bitterly incriminating charges and counter-charges that included fraud, embezzlement, and boardroom power grabs were hurled back and forth between Martinez and a longtime company executive named Richard Boyd. For the first time in its entire history, the company was subjected to a merciless spotlight, a blast of unwelcome publicity that unveiled a slew of internal tensions and threatened to pull back the curtain on the inner workings of the famously tight-lipped company. As the events unfolded, numerous questions emerged; chief among them: would In-N-Out Burger stay the same?

CHAPTER 1

The old In-N-Out Burger stand—a simple red brick structure with a twin-lane drive-through—today stands empty. Occupying a quiet lot on the corner of Francisquito and Garvey avenues in Baldwin Park, "Number One" (as it is commonly known) is surrounded by asphalt and bordered by a short row of windowless, single-story warehouses. Hemmed in on one side by the Interstate 10 freeway and a shopping complex anchored by a giant Target Center on the other, the site of the first In-N-Out is part of history now, closed off to the public by a wrought iron fence, where neat beds of white flowers grow between the bars.

At one time, on a sliver of the quarter-acre property, there was a pet shop that sold exotic birds. From the start, Harry Snyder was a man who knew how to maximize his opportunities, and oftentimes the company would lease out the unused space at various In-N-Out properties. The small shack nearly hugging the freeway ramp is still there, but it is vacant now. Its weathered green exterior is peeling and the painted lettered sign that reads "Birds ♥ Us" is still visible, although it too has been faded by time.

Directly across the street from the original store Number One is a large, empty patch of land, a circular expanse of dry earth and scrub weeds enclosed by the curving sweep of the freeway on-ramp. At one time it was home to one of Baldwin Park's most popular trailer parks.

And beyond, fanning out from Garvey Avenue, is a washed-out pink apartment building and a cluster of small ranch houses. Like the shuttered Number One, the scene is a reminder of Baldwin Park as it was before it became just another drowsy suburb of Los Angeles, when its farms and ranches were carved up, sold, and paved over in the postwar boom to make way for the eventual chain stores, strip malls, and tract houses built in their place.

But in 1948, this was the frontier of the American dream. This was where newly married couples like Harry and Esther Snyder could purchase a spit of land, buy a small house, build an even smaller burger joint—and dream big dreams. Postwar Baldwin Park was the promised land of the working class. Baldwin Park is the spot where In-N-Out Burger began. And it is where the Snyders developed their basic philosophy: serve the freshest, highest-quality burgers and fries; treat your employees well and your customers even better, all while providing friendly service in a spanking clean environment; and above all, remain family-owned and independent. That philosophy was the starting point for what would become the fast-food industry. And even as In-N-Out's competitors later raced around the globe, franchising identical box-like fast-food stores on rows of identical strip malls in nearly identical towns, In-N-Out never wavered from that simple philosophy.

Across from the old lot on Francisquito Avenue, on the south side of the freeway, stands the new model, single-lane drive-through "Number One." Opened in 2004, In-N-Out's hometown flagship is actually the third of the chain's Number Ones. In 1954, when the state of California expanded the Interstate 10, a stretch of the new freeway came straight up Garvey Avenue, cutting right through the first store—so the Snyders tore down the stand and rebuilt their second "Number One" several feet away.

Every day, hundreds of motorists whiz by the vacant "Number One" on Francisquito. They are easily confused by the low-hung In-N-Out sign out front. Many mistakenly attempt to turn into the lot before being abruptly confronted with the locked iron gate. Occasionally, however, when the gate is open and nobody is watching, it is possible to steal onto the property and glimpse the spot where In-N-Out began. Baldwin Park's "old-timers," as they like to call themselves—those folks who've lived here since that first In-N-Out opened—insist that despite its enormous success, not much has changed. "Those burgers taste exactly the same as day one," they chorus.

Looking around, it is almost possible to conjure up those early days when the area was just dirt roads and open fields and In-N-Out's dusty lot was filled with teenagers and their hot rods. Glimpsed through the large glass windows, the open kitchen remains largely intact. The single grill, despite its long retirement, is scrubbed clean and free of grease. The stainless steel beverage dispenser still has paper cups with the familiar red palm tree motif stacked up inside like Russian matryoshka dolls. And the old analog clock tower with the now famous In-N-Out logo, a bright yellow neon boomerang arrow, still keeps time as it clears the grey concrete slab of freeway that slices through the sky behind it.

That original In-N-Out Burger was once immortalized in a Jack Schmidt painting, reproduced later by the company on T-shirts and postcards; in its portrait, the squat little box of a stand with its red and white awning is offset by a blue sky filled with cumulus clouds. The snow-capped San Gabriel Mountains are in the distance, as are a handful of ranch houses. A parade of old Studebakers with their curved, single-pane windshields, sloped backsides, rear fender skirts, and fat white rim tires are lined up on the gravel drive-through lane. One motorist is shown placing an order in front of a small white box on a pole with the words "Two Way Speaker" written across the side in red letters. And manning the grill inside the open kitchen window is a smiling young man in a crisp white shirt and a paper cadet hat. There is talk that the original site depicted in that postcard will eventually be turned into an In-N-Out museum or that the old kitchen will be donated to a historical society, but nobody knows for certain.

At first glance, it can be difficult to imagine the role that this simple, unremarkable structure played in helping to reshape the contours of the American lifestyle. The fast-food drive-through has become such an integral part of our culture that it's almost impossible to imagine American life without it. Perhaps more remarkable still is that while In-N-Out began in the same plebian enclave that hatched today's monolithic fast-food industry, In-N-Out never developed into just another burger chain. At some point, In-N-Out emerged as something more compelling than a mere fast-food joint. It is nothing less than a cultural institution with a hysterically loyal cult following. Today, In-N-Out is as much a part of Southern California as the Hollywood sign or the graceful palm trees that line the streets of Los Angeles.

And yet, only the most basic of details are known about the origins and inner workings of In-N-Out Burger. The highly secretive Snyder family and their enigmatic little burger chain have always preferred to let their hamburgers speak for them. And so for some sixty years, it has stood at the curious public intersection of being both widely loved and little known. Camera shy and fiercely independent, In-N-Out has managed to preserve an inscrutable charm. Over time, that homespun charm has transformed into a hoary aura of mystique.

Yet an entire industry was influenced and an American icon was born when Harry Snyder, the gruff, self-made son of Dutch immigrants, first set eyes on Esther Johnson, a shy former schoolteacher with a gentle smile and a head for numbers. When they met, America was a different country, and Harry Snyder—an uncomplicated man with sharp instincts—had quickly sensed a hunger for a new way of doing things. The Snyders didn't invent the hamburger or even the fast-food hamburger chain; the couple discovered an entirely new market that was based on a new and changing America. Of course as is the case in many stories of American ingenuity, the Snyder's fortunes were, in part, built on the dreams of others.

At eight-thirty on a chilly grey morning in April 1906, Harry Snyder's father, Hendrick Schneider, having earlier made his way to the soggy English port of Liverpool from his native Amsterdam, joined a colorful parade of passengers streaming aboard the SS *Laurentian*. At thirty-five years old, the sturdy, brown-eyed, dark-haired Dutchman joined Jewish-Russian peddlers, Finnish farmers, Scottish tradesmen, and clusters of other Dutchmen who were all making their way to Nova Scotia.

The *Laurentian* was part of the Allan Line Royal Mail Steamship Company Ltd.; Allan's fleets, setting sail from Glasgow and Liverpool, were said to have ferried more young immigrants across the Atlantic than any other shipping company. A brochure from the era proclaimed that its steamers sailed "at fares so low as to be within the reach of almost any working man."

And so, under a gunpowder sky, the *Laurentian* edged its way down the River Mersey. Hendrick "Harry" Schneider* and his fellow passengers were no doubt filled to bursting with excitement and nervous anticipation as the ship made its way to the Irish Sea before entering the billowy, cerulean expanse of the Atlantic Ocean. As Harry's father—a painter by trade—sailed to Canada, he surely shared the dream that had spurred so many others to leave their homes, their families, and all that they had known behind them in Europe.

At the turn of the century, when Snyder *père* departed Amsterdam, he left a city in upheaval. As Schneider sailed westward, Europe at his back, Amsterdam was undergoing cataclysmic change, engulfed in riots, suffering food and housing shortages, and plagued by rampant unemployment. Schneider was just one of many European immigrants who were pouring into Canada's capacious western territories—billed as the "Last, Best, West." After landing in Nova Scotia, Hendrick did indeed head west. He moved first to Regina, Saskatchewan, before settling in Vancouver, British Columbia. In Vancouver, he found work in the shipyards as a painter.

Some three years later, in 1909, Hendrick married Mary (*née* Droewde), a thirty-two-year-old Dutch immigrant. Mary's mother reputedly came from an old wealthy Dutch family; when she married a poor diamond cutter, she was disowned, leaving Mary to grow up in something less than the comfort that her mother had known

---

* According to U.S. Census records and passenger ship manifests, Hendrick Snyder's name is spelled variously at different times as Harry, Hendrick, or Hendryck Snyder and Schneider.

as a child. She arrived in Canada with her sister and brother and (like Hendrick) they became part of a group of Dutch émigrés who lived and worked together in a close-knit community, often moving around in search of work. One of their destinations turned out to be across the border in Washington.

Over the next several years, Mary followed her husband back and forth between Vancouver and Seattle as Hendrick—who spoke broken English at best—took various painting jobs. In 1910, while living in Seattle, Mary gave birth to the couple's first child: a daughter, Lillian. Hendrick was working as a house painter, and the couple's fortunes were modest. However, by 1912, the Schneider family had once again returned to Vancouver, and Hendrick to the Vancouver shipyards. It was there that Mary gave birth to the couple's second child, a daughter they named Harriet. And a year later, on September 9, 1913, Harry Jr. was born.

When World War I broke out in 1914, Schneider had difficulty finding work in Canada. A victim of wartime prejudice, Schneider found that his name and accent were often mistaken for German, and he changed the family name to Snyder. Fearing that a German victory would further hamper his opportunities, he decided to leave Canada for good.

On September 24, 1915, when Harry *fils* was barely two years old, the Snyders sailed to Seattle. According to immigration records, Hendrick carried one hundred dollars in his pocket, a not inconsiderable sum at the time. The Snyders moved into a small house in the Fauntleroy neighborhood that faced Vashon Island and the Olympic Mountains. "By today's standards, the house wasn't much," recalled Harry many years later. "But in those days it was nice. I used to sleep in an empty chicken coop in the backyard. It was close to the beach. We could walk down there and swim or go boating right on the Sound."

The family slipped into a routine; Hendrick painted and Mary worked as a housecleaner. A census taker at the time noted that while Dutch was the couple's native tongue, both Hendrick and Mary could read and write English—in reality, the couple had learned only rudimentary English. This may be the reason that their young son Harry could not speak anything but gibberish until he was six years old.

When the Snyders settled in Seattle, it was a spit of a city in the midst of rapid expansion, nestled between the Puget Sound and Lake Washington. The gold rush of the 1890s had spurred a massive migration, and with it came huge growth. Emblematic of Seattle's progress was the Smith Tower; completed a year before the Snyders left Canada, the Tower was the tallest building west of the Mississippi River until the Space Needle eclipsed it in 1962. The Snyders' arrival coincided with the creation of many of the city's neighborhoods that began to fan out from the densely crowded downtown area. Seattle was a city of growth, offering an environment in which many tried their hand at the new, the untested, and the novel.

However, Hendrick Snyder, who could be rather volatile in temperament, appeared to be somewhat myopic in his ability to expand his own opportunities. "He had some crazy ideas," is how Harry Jr. once described his father. "He didn't believe anybody should be wealthy. He figured that if you made so much money the rest of it should go back to the government and they would give it to poor people. He was a die-hard socialist." His wife, Mary, however, exhibited few if any opinions on this subject or any other for that matter. "Ma never had much to say," Harry once explained. "She didn't make waves. He made the waves."

Despite his vocal socialist leanings, Hendrick was by his son's account something less than a visionary. Painting houses, apartments, ships—anything that needed painting—was his one undisputed talent. "I don't think you could find a better painter," Harry once said of his father. However, when it came to finances, Hendrick was careless to the point of recklessness. "The old man always ran out of money," was how Harry described his father's financial acumen—or lack thereof. "If he got paid for a job they'd have a big dinner with lots of food in the house and have a lot of people over and have a big party." During the next several years, the Snyders enjoyed living something of an exuberant hand-to-mouth existence in boomtown

Seattle. When Harry was seven years old, the circus came to Seattle. Hendrick woke him at two o'clock in the grey morning so that they could go down to watch the rail caravan arrive. "We stood and watched the animals come off," he recalled. "It was a quite a show. We watched them put the tent up. They did the show and then they left." But Harry never actually watched a performance inside the big tent. Apparently, Hendrick had already burned through his latest windfall and didn't have enough money to buy his son a ticket.

Not long after the circus left town, Hendrick left Seattle himself. Leaving his family behind for a time, he moved down to Los Angeles to find work. It was a new city for Hendrick, one in which he had a clean slate. It may have been just in time, because as the 1920s wore down, Seattle's famous boom cycle had turned to bust. And apparently Hendrick had worn out his welcome there as well. "When he left Seattle, he owed everybody and their brother money," Harry once said. "He was the biggest stiff there was."

In 1922, when Harry Snyder was nine years old, Hendrick sent for the rest of his family. The Snyders landed in a one-bedroom house on Central Avenue in the Watts section of Los Angeles followed not long after by a move to Santa Monica, a town that had benefited greatly from the expanding prosperity that was lifting most of the country during the 1920s. During this time, the Douglas Aircraft Company built a plant at Clover Field. And in 1924, the La Monica Ballroom, a fifteen thousand square foot hall capable of holding ten thousand dancers, was opened on the Santa Monica Pier. By the end of the decade, the population had more than doubled from 15,252 to 37,156.

Hendrick took a series of painting jobs and Mary continued to clean other people's houses. At almost any time, the sociable Snyders' own house was filled with an assortment of friends whom they had gathered from the Dutch immigrant community. While good times were readily available, money remained in short supply. As a result, the family moved from one rental house to the next, often owing rent to the landlords they had left behind. When Harry was thirteen, his father landed in jail after he beat up a landlord who came by demanding his money.

The Snyders moved along the Los Angeles beach circuit. In addition to living in Santa Monica, they stayed at a series of small houses in Venice. Founded in 1905 by tobacco magnate Abbot Kinney as a beach resort, Venice was considered the Coney Island of the West with its pleasure piers and amusement parks, speakeasies, dance halls, miniature railroads, bathhouses, and ocean side canals complete with gondola rides. Venice also became associated with the "Craftsman" style of architecture. At one point in 1930, the Snyder family lived in a single-family Craftsman home at 221 Market Street.

In 1928, Hendrick Snyder quit his job painting at the three-story, sixty-bed Santa Monica Hospital on Sixteenth Street. The work paid a tidy sum of twenty-five dollars per week, but when Hendrick was refused a vacation, he simply put down his paintbrush and walked away, leaving his wife to support the family with her housekeeping jobs. The Great Depression plunged the country into an economic morass a few years later, and Hendrick never held another proper job again.

These were tough times. During young Harry's teenage years he was a bit of a tough himself, engaging in petty theft and sparring with his friends at the amateur boxing gym in Ocean Park. Throughout, he managed to maintain a fierce sense of responsibility. Although he was just a young man and the Depression had flattened prospects for most, Harry always found a way to earn a buck. He landed a number of odd jobs—and he always supported his family, giving them five dollars every week from whatever money he had earned.

Among his many jobs, Harry worked as a paperboy. Delivering the *Venice Vanguard* on his bicycle, he boasted that he could finish his route up and down Washington Boulevard in half an hour. Among his other jobs, Harry worked for a grocer, sold sandwiches, and delivered hot dog and hamburger buns for a bakery. And for a period he worked at the concession stands on the pleasure piers of Venice. Even as the cheap entertainments offered on the piers grew increasingly shoddy, attracting a more vulgar crowd to the once up-scale beaches, Harry picked up a number of odd jobs doing everything from selling red hots to picking up rental umbrellas on the beach. "It was not very much money," he once remarked. "But money in those days was money."

By his own admission, Harry was not much of an academic. Despite his innate intellect, he was never more than a middling student. "I went through high school with a C-average without cracking a book at all," he once said. "A teacher told me if I had ever studied I could've gotten good grades." Rather, Harry excelled in the areas of common sense and resourcefulness. An amateur electronics enthusiast, he demonstrated the kind of mind suited to putting together and taking apart gadgets.

During his high school days, Harry developed a keen interest in cars and a fondness for smoking Chesterfield cigarettes—a habit he picked up while playing a regular game of rummy with his buddies. Following high school, he enrolled at Santa Monica Junior College, but quit after one semester. "I couldn't afford to go," he later explained. "I had to work to keep myself going." Many years later when Harry looked back at this period he claimed that the times had briefly transformed him into a radical. "I was a communist," he declared. "I saw the capitalist system as a total failure. We needed a whole new system, and communism was the only answer. Everyone working together and sharing it all instead of some getting it all." In all likelihood, Harry's drive and tenacity were propelled by the uncertainty of watching his parents labor to provide for his family, chasing one opportunity after another, and by the Depression years. And despite his youthful transgressions, Harry grew into a disciplined fellow with a strong sense of responsibility. He had learned early that luck was not something that found you; if you wanted it, you had to go looking for it yourself.

By the time the United States entered World War II in 1941, Harry had already begun to make some luck of his own. Having returned to Seattle at one point, he worked as a railway signal operator. During the war, Seattle became a center for industrial manufacturing; it also served as a major embarkation point for troops. And at age twenty-nine, Harry Snyder was drafted. On November 23, 1942, he walked into an army recruitment office in Tacoma and signed up for duty.

Harry was quickly sent to basic training in Northern California at the Fresno County Fairgrounds that had until only recently been used as a crude temporary camp holding 5,344 ethnic Japanese residents. By the time that Harry Snyder arrived, the last of the Japanese had been transferred and the fairgrounds turned over to the Fourth Air Force. Although no less crude than when it was a Japanese internment camp (makeshift tar paper barracks, outhouses, and overhead water pipes with drilled-out holes), the Air Force converted the grounds into a non-flying training facility for signalmen, camouflage specialists, chemical warfare specialists, clerks, cooks, and truck drivers.

A perforated eardrum disqualified Harry Snyder from the infantry and saved him from certain deployment overseas. Instead, Harry's service was largely performed behind a desk. Following his basic training in Fresno he was stationed variously in Hastings, Nebraska; Dallas, Texas; and Salt Lake City, Utah. For a time, Harry was sent to Hamilton Field in Novato, California, northeast of San Francisco. It was the headquarters for the First Wing of the Army Air Corps, where he worked in the records department processing B-24s. For extra cash, Harry worked on the side in the Sausalito shipyards, earning an additional seventy dollars a week. Following his tour of duty at Hamilton Field, Harry was sent back to Los Angeles, where he worked as a clerk-typist for the Army Air Corps.

When the war ended, Harry returned to Seattle. The war years had shuffled people far from their homes and thrown them together in the most unexpected ways. In 1947, Harry was thirty-four years old and working as a caterer selling boxed sandwiches to the cafeteria at Fort Lawton. And one day in 1947, while dropping off his sandwiches at Fort Lawton, Harry met the restaurant's twenty-seven-year-old manager, a shy woman with a warm, engaging face. Her name was Esther Johnson.

The daughter and granddaughter of coal miners, Esther Lavelle Johnson was born on January 7, 1920. She grew up in the tiny village of Sorento among the vast rural plains of coal-rich Shoal Creek Township in southwestern Illinois, the fourth child of Orla and Mabel (*née* Molloy) Johnson's brood of eight.

When Esther was a young girl, the family lived in a small, one-story wooden frame house with a shingled roof on State Street, across from the United Methodist Church. Sorento was so small that most houses—including the Johnsons'—didn't have numbers. The cramped house was perhaps not more than one thousand square feet; it had a small front porch and a tiny yard that pressed up to the street. The house was unexceptional, but it was certainly a step up from the miners' camps where many families in neighboring counties lived.

A fifth-generation Illinoisan, Esther hailed from a long line of hardy men and women who possessed a spirit of faith and great pluck. On Esther's paternal line, her great-great-great-great-great-grandfather Christopher Loving was born in Virginia in 1750 and fought in the Revolutionary War as a teenager. A thrifty man who kept scrupulous accounts of his money, Loving owned a hundred-acre farm in Chester County, South Carolina. James Grisham, Esther's paternal great-great-grandfather was a farmer and a veteran of the great Black Hawk War of 1832. Originally from Dixon County, Tennessee, he was one of the first to settle Montgomery County, Illinois.

Narrowly escaping the great Irish potato famine, Esther's maternal grandfather, Brien Molloy, was born on March 1, 1850, in New Orleans, just six weeks after his parents, Patrick and Catherine (*née* Monahan), arrived from County Cavan, Ireland. Later, Molloy traveled north to Illinois where he became one of Sorento's early businessmen. He and his wife, Anna, had eleven children including Mabel, Esther's mother.

The Johnson line of the family came to Sorento with the arrival of Esther's great-grandfather Israel Johnson. A farmer and auctioneer, Israel was born around 1821 and traveled to Illinois by covered wagon from Tennessee. According to a census taker, in 1870, Israel Johnson owned a personal estate worth $1,000. Israel's father, Jonathan Johnson, was a circuit-riding minister from Tennessee. In those days, the circuit riders traveled on horseback for weeks at a time with all of their earthly possessions packed into a saddlebag, going from provisional town to provisional town and makeshift church to makeshift church, taming both the wilderness and those lost souls that they passed along the way.

At one time, Esther's mother, Mabel, worked as a schoolteacher. Before his marriage, her father, Orla, worked in a retail store. However, like his father, Lawson, before him, Orla eventually became a coal miner and a farmer. It was difficult for most men of modest means not to go down into the mines. Cheap bituminous or soft coal was plentiful in Illinois, and by the mid-1800s it was in great demand. Until the onset of the Great Depression, coal output from Sorento soared.

Orla, like many locals, ended up working at the Shoal Creek Company's Panama Mine. First opened in 1906, the Panama was just five miles from Sorento. Orla worked alongside his brother Olis, their father, Lawson, and a man named John L. Lewis (who went on to serve as the president of the United Mine Workers of America for forty years, from 1920 to 1960).

The Panama was a relatively small operation, used primarily to supply the Clover Leaf Railroad (built in 1881 along the south shore of the Great Lakes, connecting Buffalo, New York, and Chicago). The mine had a history of closing down for long periods, and so most of its workers supplemented their mining income with other work, mostly farming.

By the time Esther's father began working the mines, coal mining was already on a downward slide. The stock market crash in 1929 followed by the Great Depression ended the mining boom for good. Thousands of coal miners lost their jobs as mine after mine was shut down. In 1934, when Esther was fourteen years old, the Panama shut down permanently. Groups of miners and their families emptied out of Sorento as they went looking for livelihoods elsewhere.

While the Johnson family remained in Sorento, they were not spared hardship. At one point, the family farmed out some of their children to live with relatives. Esther was sent for a time to live with her grandparents. Although it was meant to be a temporary measure, she ended up living with them until she finished high school.

Although her parents were Catholics, while living with her grand-parents, Esther regularly worshipped at the Free Methodist Church. A simple, whitewashed, wood plank frame structure with pine pews and a shingled bell tower, the church was located on the west side of north Main Street across from the railroad tracks. The church took its name from its break with the practice of charging for better seats in the pews closest to the pulpit (Free Methodists were also opposed to slavery). It was here that Esther's own simple and principled beliefs were most likely formed.

Whatever aspirations the Johnsons may have had for their eight children, the modest circumstances in which they found themselves certainly narrowed the possibilities. Yet despite the financial con-straints that bore heavily on the family, they managed to place an emphasis on education. By all accounts, Esther was a gifted student; she possessed a sharp brain disguised by a shy demeanor. In her high school yearbook, the *Sho-La-Hi*, Esther is shown wearing a sensible, modest dress and thick eyeglasses that make her look much older than the demure teenager she was. During Esther's three years at Sorento High School, she was an honor student as well as a member of the glee club, the school chorus, and the commercial club. She was also class president and assistant editor of the *Sho-La-Hi*. With char-acteristic modesty, Esther once described her school days by saying that "I had a good time in school."

Like many graduates of Sorento High School in the late 1930s who wished to attend their fourth and final year of school (Sorento only offered three years), Esther had to continue at one of the high schools in either Hillsboro or Greenville for a fee. Although it's not clear how she managed it or who paid the bill at the time, Esther did go on to finish and graduate from Greenville High School in 1937, commuting sixteen miles every day. Esther's Greenville yearbook in-scription reads, "If it takes brains, she'll get along."

Esther's desire to attend college was stalled when her grandmoth-er suffered a stroke and was left unable to walk. Esther remained in Sorento in order to take care of her grandmother, putting off her studies until 1939, when she enrolled in Greenville College (part of the Free Methodist Church). An aunt paid for her first two years of tuition. "She thought I deserved it," Esther later explained. But her grandfather and uncles seemed perplexed as to why a young lady like Esther wanted to pursue a college education. "Don't you want to get married?" they regularly asked her. To which she replied: "I like to learn." During the mornings, Esther cared for her grandmother. In the afternoons, she studied elementary school education at Greenville; upon returning home in the evenings, she took care of the household chores.

In November 1940, Esther's grandmother sickened; Esther quit college altogether to look after her grandmother full-time. Four months later, after the death of her grandmother, Esther spent the summer in Sorento working a few jobs, including one at the post office. She borrowed $150 and returned to Greenville College. After earning the equivalent of a teaching certificate, Esther began teaching second grade at Sorento's primary school.

It was while Esther was working as a schoolteacher in her hometown that the United States entered World War II. In 1943, Esther joined the war effort, signing up as one of the navy's newly created WAVES (Women Accepted for Voluntary Emergency Service) at the recruitment office down in St. Louis. It was a bold move for a shy girl from rural Illinois. The WAVES were designated by special order of President Franklin D. Roosevelt to fill the vacancies created by the tens of thousands of men who had been sent to the battlefront. While not eligible for combat duty, WAVES performed a number of jobs previously uncharacteristic for women in the military. In explaining her reasons for joining the navy, Esther remarked, "I thought I might enjoy radio or hospital work. My parents didn't really comment on it. But my sisters thought it was wonderful."

Esther—who had very little experience outside of Sorento before enlisting—was sent to New York City, where she attended boot camp at Hunter College in the Bronx. By special war decree, Hunter was transformed from a college to a naval training ground. Once there, the new recruits spent several weeks immersed in naval organization, administration, and history. In addition, the women were subjected to an intensive program of marching drills and medical exams as well as a battery of aptitude tests. Following boot camp, they were dispatched to their duty stations. Esther was sent to the San Diego Naval Hospital; the navy proved to be her ticket out of the farms and coal mines and the chronic privations of Bond County, Illinois—just as it was for the many young men across the country that had enlisted in droves.

For the first time, the war put women on an equal footing with men, if only temporarily. WAVES recruiting posters from the period promoted the idea of women contributing to the war effort as central players and not merely as support staff to the men. They depicted females with expressions of serious purpose. They had carefully coiffed hair, wore smart uniforms, and performed such roles as air traffic controllers and parachute riggers. Underscoring the posters' imagery were their taglines. "Share the Deed of Victory," said one. "Don't Miss Your Great Opportunity," implored another.

In her official navy portrait, Esther appears transformed from the mousy, shy girl of her student years to a picture of 1940s glamour and sophistication—not unlike one of the WAVES in those navy recruitment posters. Gone are the glasses and the slumped shoulders of her high school yearbook. Instead, Esther wears an expression of self-assuredness. She is smiling brightly and her lips are painted a dark shade of crimson. Under Esther's smart navy cap, her face is framed by a short, stylish bob of curls. It seems that in the navy, Esther blossomed.

Esther Johnson spent the next three years (until November 1945) in the navy. Among her duties, she served as a surgical nurse and a pharmacist's mate. Her time visiting with burn victims left a lasting impression—it was the reason that Esther became a lifelong supporter of veterans' causes. When the war ended and Esther left the navy, she had earned the rank of pharmacist's mate second class.

Following the war and her discharge, Esther returned briefly to Sorento to visit with her family. During her service, Esther had become enchanted by the West Coast, and following the war it was her desire to resume her college studies. A like-minded girlfriend had some friends in Seattle, and the pair decided to head to Washington.

Esther enrolled at Seattle Pacific University, earning a degree in zoology while working the night shift at a laundry and later becoming a manager at the restaurant at Fort Lawton.

One day in September 1947, while working her shift at the restaurant, she caught the eye of Harry Snyder. "He just came in to deliver sandwich boxes," she remembered. "Boxes and boxes of sandwiches." At thirty-four, Harry was seven years Esther's senior and a head taller than her. He was a sturdy man with a long face, a broad nose, and large basset hound eyes. A no-nonsense kind of fellow, he was not a particularly tall man or classically handsome, but he carried himself well. "He had real friendly eyes," Esther later recalled, "full of energy." Although she found him "nosey" at first— "he was asking all these questions and found out about your life history in no time," she said—Esther soon warmed to him.

The pair seemed to complement each other perfectly. While Harry was tough and could be demanding, Esther, widely regarded as a gentle soul, did much to soften Harry's rougher edges. He had street sense; she was book smart. Where he was a maverick, a visionary, she was grounded, practical. Harry was hard-nosed; Esther was sentimental. He was a forceful presence, while she preferred to get things done quietly, behind the scenes. They were both kindhearted souls who shared a generosity of spirit. Harry believed in his ability to create his own opportunities—and Esther believed in Harry.

After a brief courtship, the pair married in 1948. They decided to leave Seattle and start their life together in Southern California. Harry, who grew up there and whose family now owned a bakery, arrived first; Esther joined him a few months later. At Fort Lawton, Harry had developed take-out bagged lunches to serve the scores of soldiers passing through and he had come up with an idea for a new kind of restaurant. America was on the verge of entering a new age, and Southern California was lighting the way. And so, just as several previous generations of the Johnson and Snyder families had pressed forward, pulling up stakes, chasing down the promise of a brighter future, Esther and Harry were starting over—together.

CHAPTER 2

The new couple set their sights on a rural town called Baldwin Park, seventeen miles east of downtown Los Angeles, in the heart of the picturesque San Gabriel Valley. When the Snyders arrived, Baldwin Park looked much as it had when it was pasture land belonging to the San Gabriel Mission. Some seven thousand people lived within the town's pastoral 6.8-square-mile stretch, nestled among acres of orange groves and citrus orchards that were still protected from the frost by black smudge pots. Largely undeveloped, Baldwin Park was primarily made up of small farms, vineyards, chicken and cattle ranches, and dairy farms. Rows of eucalyptus, black walnut, and pepper trees shaded its dusty roads like army regiments that gave way to open fields crowned by blue skies, ringed by the improbably snow-capped San Gabriel Mountains, and bathed in the rays of the sun.

Despite Baldwin Park's rural American setting, its residents proudly claimed two symbols of modernity. Leafy Morgan Park with its lighted baseball field was built before the Depression, costing some $28,000. Perhaps more significantly, Baldwin Park could boast that it was on the new Pacific Electric Railway line. Built in 1902, the Pacific Electric was the largest electric interurban system in the world at the time. Running sixty trains daily between San Bernardino and Los Angeles, the Red Line ferried passengers from the Baldwin Park depot to downtown Los Angeles in thirty-five minutes flat. Before In-

N-Out Burger, the biggest name in Baldwin Park was the traveling Al G. Barnes Circus. With no fewer than twelve hundred performing animals, Barnes billed itself as the largest wild animal circus in the world. In 1927, the flamboyant Al Barnes snapped up a quarter-mile tract of land facing Valley Boulevard for roughly $200,000 for use as the circus's winter camp after he was forced to leave his previous encampment in Culver City (where locals had grown wary of boisterous circus employees, who reportedly bought liquor from bootleggers—homeowners also complained that during feeding time, they could hear the roar of lions a mile away).

The residents of Baldwin Park happily adopted the circus and its numerous creatures. During a building binge, circus elephants were used to pull down trees and haul away timber later used to build homes. Indeed, the circus gave the town a measure of fame, attracting some of Hollywood's tinsel when the Paramount movie studio used the Barnes lot to film *King of the Jungle* with Buster Crabbe. Delighted residents were used as extras. The town's thrilling circus days came to an end in 1938 when the show was finally shut down not long after Ringling Brothers purchased it.

When Harry and his new bride arrived, the couple found an unexceptional town of wide-open spaces filled with restless teens. A bucolic stretch on the citrus belt, Baldwin Park had only received dial-up telephone lines three years earlier. In 1948, Baldwin Park was on the verge of transmuting into yet another car-obsessed suburb of Los Angeles. Like the rest of Southern California, Baldwin Park was undergoing a rapid transformation fueled almost exclusively by postwar development.

World War II was already receding into the past. New borders were being drawn and a new geopolitical landscape was being defined. The British Empire was in its twilight years; India and Pakistan were independent from both the United Kingdom and from each other. And the United States had quickly stepped out from behind England, establishing itself as the strongest, most influential, and most productive nation in the world. In something of a surprise, Harry S. Truman was reelected president of the United States, soundly defeating Republican nominee Thomas E. Dewey. And Truman was Baldwin Park's kind of fellow. Its voters, reported the *Baldwin Park Tribune*, had "upset Republican hopes, confound[ed] Mr. Gallup and surprise[d] even members of the Democratic party." An undisputed optimism was sweeping the land, and nowhere was that more obvious than in Southern California.

Lured by the promise of opportunity, a mild climate, and relatively inexpensive real estate hitched to an enormous economic rocket, the Snyders—like hundreds of thousands of other newcomers—moved to the Southland in record numbers. Los Angeles had earned a reputation as something of a flourishing boomtown, benefitting greatly from the fact that it was situated on one of the most verdant agricultural belts in the country. And throughout much of the 1920s and 1930s, the Southland enjoyed a prosperity buoyed by the oil and movie industries.

During the early 1940s, California played a tremendous role in the country's war effort. The military unfurled its industrial flags across the state; defense contractors established huge manufacturing centers, the aerospace industry set up shop, factories were built, and military bases were erected. Between 1939 and 1945, federal spending reached more than $35 billion in California, making it the third largest manufacturing center in the country behind New York and Michigan. The ripple effect was enormous. Waves of growth radiated out from Los Angeles like a fan. New jobs were created, new development and infrastructure followed the jobs, and an influx of new residents trailed closely behind; the average personal income tripled. As a result, throughout the war years, some 1.6 million Americans migrated to California. Many received their military training here, while others touched down on their way to the frontlines of the Pacific theater. Following V-Day, a number of those who had gotten a taste of the Golden State decided they didn't want to leave.

It was nothing less than a mass exodus of Americans. Indeed, between 1940 and 1950, the state population swelled from 6.9 million to 10.5 million, an increase of approximately 52 percent. Between July 1945 and July 1947, more than a million people resettled in California—many ended up in the Southland. In the words of writer and social commentator Sarah Comstock, "As New York is the melting pot for the peoples of Europe, so Los Angeles is the melting pot for the peoples of the United States."

Baldwin Park quickly became one of the fastest-growing postwar communities in this vast and growing melting pot. By the time the Snyders arrived, the number of inhabitants had nearly doubled from its prewar population (by 1956 it would nearly quadruple to 28,056). In 1948, after the local post office reported an all-time high of $65,000 worth of postal receipts, the *Baldwin Park Tribune* concluded, "Since the business done by the post office is considered a reliable weather vane of community progress, the receipts for 1948 can be taken as an indication of a doubling of the importance of the community during the past ten years." A less strictly numerical (although no less noteworthy) indicator of Baldwin Park's growth was the fact that the town now boasted eighteen different churches, all supporting their own congregations of Lutherans, Episcopalians, Methodists, and Catholics, among others.

The surge of people and activity in Baldwin Park led to a resumption of the growth and prosperity that had been paralyzed by the Depression. A number of the large ranches had been broken up and subdivided into smaller lots and sold at bargain prices. While the scent of citrus had always filled the air in the San Gabriel Valley, now, for the first time since before the stock market crash of 1929, the air was thick with new possibilities.

New residents were pouring into Baldwin Park at a rate that outpaced available housing. Many families ended up living in the numerous trailer parks that had emerged in the wake of a significant housing shortage. One of the largest, Baldy View, had been built on a circular piece of property on the northeastern corner of Garvey Avenue, alongside Johnny No-Bones Steakhouse #1, with a permit to operate 180 camp spaces. The housing boom attracted numerous speculators and developers; they bought up large tracts of land, turning the

San Gabriel Valley (along with the San Fernando Valley) into what historian and author Mike Davis called a "real estate casino."

For the average American, home ownership seemed an unobtainable dream before the war. But it was well within the grasp of many during the postwar boom, partially as a result of the generous GI Bill that offered $2,000 home loan guaranties. Indeed, between 1944 and 1952, the Veterans Administration backed nearly $2.4 million in home loans for World War II veterans. In Baldwin Park, shortly after the war, a desirable two-bedroom house with a double lot cost between $6,650 and $8,500. And a large percentage of the town's new housing developments were specifically built with the influx of veterans in mind.

One residential builder, Louis Rudnick, developed a tract of two-bedroom houses with the lofty sounding name "the Baldwin Park Estates." They were generally referred to simply as "GI homes," and were made available to veterans "with no down payment except escrow and impound costs." The houses boasted such features as automatic garbage disposals and plenty of closet and storage space. Another developer, Home Builders Institute, constructed a tract known as the Baldwin Park Homes. They were two- and three-bedroom dwellings along Los Angeles Street at Merced Street, between Las Tunas Drive and Ramona Boulevard, and were offered to veterans for "as low as $250 down." Baldwin Park's rural landscape of orchards, dairy farms, ranches, and citrus groves was fast being transformed into stretches of ranch houses and cottages with detached garages.

The Snyders moved into a small one-bedroom house on the corner of Vineland and Francisquito avenues, across the street from the Baldy View trailer park. It was a time marked by uncomplicated formality, of starched white shirts for men and peplum skirts for women, and of hopefulness and sanguinity for all. The era carried all of the hallmarks that came to be associated with Southern California—a place where the past could be cleared away in order to build a brighter future. Indeed it was during this period that a Chicago-transplant, cartoonist, and animator named Walt Disney bought up 160 acres of orange groves and walnut trees in Anaheim and erected an homage to his dreams that he named Disneyland.

Harry Snyder's dream was a modest one. He was going to start his own little food business—a hamburger stand. Harry had good reason to want to go into business for himself; a child of the Great Depression, he watched his father shift between the Vancouver ship-yards and the boom-bust cycle of Seattle in search of work before moving his family down to Southern California in search of yet another opportunity, only to land in the middle of the greatest economic decline in American history. Harry was determined to live a different sort of life. It wasn't greed that drove him; he was simply determined to create his own future.

With the war over, Harry had a strong gut feeling. Although he had a couple of failed businesses under his belt, he believed that following the Depression and years of war and shortages, people would be in the mood to enjoy life. Harry wanted to serve quality food at a reasonable price, and as quickly as possible. "We really have to have a place where people can get their sandwiches and go," he said. Harry and Esther would open a new kind of hamburger stand—the drive-through—catering to an increasingly mobile society. A typical entrepreneur, Harry was rich in ideas but short on cash. Early on, he took on a partner, a man named Charles Noddin, who agreed to finance the venture. Noddin reportedly put up some $5,000 in start-up capital.

Harry Snyder's instinct was a good one. Southern California was the most heavily motorized place on the planet. Already by 1940 there were over 1 million cars in Los Angeles—five automobiles for every four families. It was reported at the time that Angelinos spent more money on their cars than their clothes. Moreover, the Los Angeles Police Department had more traffic officers than the San Francisco Police Department had policemen on their entire force.

In the wake of World War II, small food stands selling burgers, tamales, and hot dogs were spreading across the Southland like wildfire. They were—at least in the beginning—little more than shacks fronted by a couple of stools, needing only a minimal amount of space.

The legendary Original Tommy's World Famous Hamburgers got its start when Oklahoma City–born Tommy Koulax, the son of Greek immigrants, began selling his twelve-ounce chili cheeseburgers out of an eight-foot by fifteen-foot cinderblock stand on the corner of Beverly and Rampart boulevards in downtown Los Angeles. Koulax, a shipyard welder during World War II, initially used his 1929 Model A Ford as his storeroom. On May 15, 1946, Tommy's first day of business, Koulax drummed up eight dollars in sales. "My grandmother told me he spent so much time there that he slept on top of the onion sacks," recalled Koulax's granddaughter Dawna Bernal. So busy was the "Shack," as it was affectionately known, that in 1949, Koulax turned Tommy's into a twenty-four-hour stand, enlarging the tiny space by adding an awning. In 1965, Koulax was able to purchase the entire southeast corner of Beverly and Rampart (eventually he came to own three of the four corners). And in 1970, Koulax boasted that Tommy's had raked in $1 million; soon he began to expand his popular chain across Los Angeles. When Koulax died in 1992 at the age of seventy-three, there were seventeen Tommy stands and the original was selling twenty-five thousand burgers each week.

Like Tommy's, many of the stands were fairly lucrative. There was a demand for cheap food from a constant flow of customers who worked in nearby factories, mills, offices, and shops. Not coincidentally, this casual new way of dining dovetailed with the rise of car culture and the establishment of the extensive interstate highway system that was starting to crisscross the nation. With better roads, people could travel farther. Along the way people would need places to rest, sleep, and above all eat.

These little burger stands heralded a new way of eating. And this new way of eating eventually had an impact on America's economy, topography, popular culture, and even the way Americans were viewed abroad. In time, there were negative consequences, to be sure, but in the period immediately following World War II, these little burger stands signaled sweeping changes in the country. And the capital of this grand sweep of changes was Southern California, where love for the automobile was perhaps only rivaled by love for the hamburger.

CHAPTER 3

Short on experience but long on common sense, Harry sought advice from Carl N. Karcher, one of fast food's pioneers, who had built a small, growing chain of hot dog stands in Los Angeles called Carl's Jr. A sharecropper's son from Upper Sandusky, Ohio, Karcher was an ebullient, salt-of-the-earth character who would go on to build a fortune transforming his handful of hot dog stands into the $1.5 billion, 3,000-unit, multinational Carl Karcher Enterprises. Looking back, Karcher said that he wasn't surprised that the Snyders decided to seek his counsel. "They came to see me because Harry saw a successful business," he explained self-assuredly. "A successful fast-food business. He didn't go to see someone in medical sales."

An eighth-grade dropout, Karcher was working long hours on his family's Ohio farm, milking and feeding cows, when in 1937 an uncle asked young Carl to join him at his Feed and Seed store in Anaheim, California. After careful deliberation, the twenty-year-old picked up and drove over two thousand miles to Anaheim where he worked selling chicken and cattle feed to local farmers for eighteen dollars per week.

A devout Catholic, Karcher met his future wife, a local girl named Margaret Heinz, at Anaheim's St. Boniface Church one Sunday. Karcher said he was instantly smitten with Margaret, who was living with her parents and fourteen brothers and sisters in a two-story Spanish-style home surrounded by orange groves and chickens. However, despite his infatuation with Margaret, Karcher briefly returned to Ohio. "I wasn't sure about California," he exclaimed, "but I missed Margaret." So he turned around and headed back to Anaheim.

On November 30, 1939, not long after Karcher returned, the couple was married. However, his uncle had refused to rehire his nephew, and Karcher went to work for the Armstrong Bakery on Avalon Boulevard wrapping and delivering bread. "He said anyone that quit he wouldn't hire back," explained Karcher. Waking up each morning in the hazy light of dawn, Karcher drove his truck delivering bread to local restaurants. During his rounds, he noticed the increasing number of hot dog stands that were popping up all over the city and quickly calculated the number of buns he was unloading on a weekly basis. "I went to about half a dozen of the hot dog carts twice a day," he recalled. In the summer of 1941, when the cart on Florence Avenue became available, Karcher snapped it up. "I thought it was a good opportunity to own it," he said. Karcher secured a loan from the Bank of America using his Plymouth Super Deluxe as collateral and another fifteen dollars that he fished out of his skeptical wife's purse to purchase his first cart, located across the street from the Goodyear factory. "I was concerned at first," Margaret Karcher later explained. "Back then, you didn't think of going into debt, and what were we going to do? We didn't have the money to buy a hot dog cart." But Karcher was determined and the transaction was completed. When he showed his wife the handwritten receipt signed "received from Louis Richmond for $326 for cart 7/17/41," she dryly told her husband. "I guess we have a hot dog cart."

Margaret Karcher's doubts evaporated quickly. The little stand did a brisk business selling hot dogs, chilidogs, and tamales for a dime apiece and soda for a nickel to the plant employees working double shifts during World War II. More than six decades later, Karcher proudly recalled his take on his first day of business: "It was July 17, 1941. We made $14.75."

A year later, the Karchers purchased a second cart; Carl worked at one while his wife manned the other. When Karcher was drafted in 1945 (he served as a cook at Fort Ord in Monterey Bay, California), Margaret took over and ran the business. When the war ended eight months later, Karcher returned from his military service to find that the small business he had left behind was thriving.

By the time the Snyders called upon Carl Karcher in 1948, he had four hot dog stands and had opened his first full-service restaurant called Carl's Drive-In Barbeque in Anaheim, down the street from Margaret's family home on North Palm Street (the property would later become the first headquarters of Carl Karcher Enterprises). Less than ten years after leaving his family farm in Ohio, Karcher had reached a level of success that stretched beyond even his own imagination. "I always heard that if you had three or more locations you were a chain operator," he later said.

At thirty-one years old, Karcher was a few years younger than Harry Snyder—but he was an elder statesman when it came to the fast-food business. That first meeting would be the start of a lifelong friendship between the Karchers and the Snyders, who went on to become friendly rivals. The two couples had much in common. Tireless and hands-on, each pair worked as a team, and both couples went on to build their businesses based on equal doses of hard work, strong faith, and dedication to family (the Karchers eventually had twelve children). Karcher's hardscrabble childhood had produced the belief that luck was something you made yourself, a lesson with which Harry Snyder was all too familiar himself.

When Karcher met the Snyders, Carl told them to focus on a great product and to maintain the personal touch. "It's so important to make people feel special," he explained. Karcher also shared his core value. "My whole philosophy is never give up." However, during that first meeting, Karcher found that the aspiring entrepreneurs already had very specific ideas about how they planned to run their business and treat their employees. Many decades later, when he was "ninety-years young," Karcher still appeared struck by it. "They were very particular about their people smiling," he said. "They wanted their employees to feel like they were part of the company, like they were owners themselves."

Despite suffering from the advanced stages of Parkinson's disease, Karcher could still recall his first impression of Harry and Esther Snyder. "They were two great people who were dedicated to each other," he said. Affable and surprisingly munificent, Karcher explained that at the time he had no problem sharing his expertise with the couple. "I have always said that competition just makes you stronger. You shouldn't be afraid of the competition. They make you stay on top of your game. If another chain started near a Carl's Jr., it showed that we had a good location and it brought more people in. I told the Snyders that it's very important to have respect for your competitors. I may have had a different philosophy than some of the others. But I believe that your competitors are really your friends. They keep you on your toes."

On October 22, 1948, the Snyders opened their first burger joint across the street from their house on Garvey Avenue. It was a modest, low-slung box on a very small lot. It was Harry's philosophy to stress fresh ingredients and high quality. As his nephew Bob Meserve (who began working for his aunt and uncle in 1962, when he was seventeen years old) explained, Harry "wanted to take the lettuce out of the ground, the tomato off the vine, and the onion and prepare the burger fresh right now. That was his goal."

The couple sold a spartan menu of twenty-five-cent hamburgers, thirty-cent cheeseburgers, fifteen-cent french fries, and ten-cent soft drinks. Aside from a tiny nearby grocery, Nordling's Market, and the large trailer park directly across the road on Garvey, the neighborhood was marked by wide swaths of land crossed by dirt roads and the San Gabriel Mountains in the distance. The stand was situated on the road to the neighboring town of El Monte; Baldwin Park itself was located on the well-traveled route between Los Angeles and Palm Springs.

It was Harry who came up with the new venture's name—IN-N-OUT HAMBURGERS—and the couple erected a red neon sign with elongated corners, white block lettering, and stripes. The sign, pitched by the side of the road, broadcast In-N-Out's early motto: "NO DELAY."

Harry had anticipated the significant role that the car would continue to play in California. Baldwin Park was one of many small towns linked by streetcars that were fast transforming themselves into tract houses connected by a network of high-speed highways. The highways soon replaced the streetcars altogether. American life was becoming increasingly mobile. The exodus from the cities in favor of the suburbs meant that people had longer commutes. More women were working and less and less time was devoted to food preparation in the home. One of the first casualties of the new on-the-go lifestyle was the sit-down meal.

These little restaurants cropping up all over on the roadsides were known simply as the drive-in. Many consider the Pig Stand, which opened on the corner of Sunset and Vermont in Hollywood in 1932, to be the first. It got its start in Texas in 1921 by Dallas candy and tobacco magnate Jessie G. Kirby, who appreciated the growing attachment people had to their cars, and his partner Reuben W. Jackson, a prominent physician. The early Texas Pig Stands used waiters who took orders and served customers still parked in their cars. By the time the chain had grown to sixty locations, carhop drive-ins were popping up all over.

Drive-ins became an adventure in eating. It was said that the number of drive-in restaurants rivaled the number of automobiles in Southern California. Food stands with bright neon signs and carhop girls, those attractive young women dressed up in flashy costumes who served patrons in their cars using specially made trays attached to car windows, dotted the landscape. In fact, they came to define the landscape. Popular in Southern California, they soon spread to the rest of the country. "Houston Drive-In Trade Gets Girl Show with Its Hamburgers," was the headline accompanying a 1940 *Life* magazine cover story featuring Sivil's Drive-In in Texas—the story featured a full-length photograph of a Sivil's carhop outfitted head-to-toe in her satin majorette costume.

The surge in motorists and motoring meant that there was a

captive audience of potential customers. And to attract them, the drive-ins used architecture, the kitschier and louder the better. Square and rectangular buildings became circular neon palaces; fanciful colored lights swirled and flickered, inviting passersby. Nothing was too whimsical or outlandish. There were drive-ins in the shape of outsized hot dogs, bowls of chili, donuts, and even gigantic root beer bottles. On the California border with Tijuana, the Gorro Drive Inn was built in the shape of a giant sombrero. And Seattle's well-known Igloo Drive-In had two domed igloos that could seat seventy inside while carhops dressed in ski-togs and white boots in the winter and short skirts in the summer served "Husky Burgers" and "Boeing Bombers" in the parking lot.

Then In-N-Out arrived.

The Snyders' burger shack was tiny, it had no indoor seating, and there was little room for a full-fledged drive-in with carhops. Harry, an amateur electronics enthusiast, came up with an idea that would compensate for these deficits. He dispensed with the carhops altogether and replaced them with an invention of his own: a simple two-way speaker box made out of a few off-the-shelf electrical components that was connected to the eatery's kitchen. That way, motorists could order at one end of a driveway into a small white box attached to a pole dug into the road and then proceed over the gravel drive-through lane to pick up their food at the other end.

That's how Harry ended up with the name In-N-Out Hamburgers. The simple, to-the-point name reflected the uncomplicated view of the Snyders. Later abbreviated to In-N-Out Burgers, it was lucid shorthand for this new model burger joint: customers driving *in* to order their food and then driving *out* without leaving their cars. Harry's two-way speaker box was an invention that was as innovative as it was practical; it perfectly reflected America's growing interest in technological gadgetry and its growing obsession with rapid consumption and carryout food. In-N-Out Burger was just the right kind of eatery for a new kind of America, an America that was constantly on the go.

Yet, Harry's new speakerphone drive-through format got off to something of a troubled start. In 1948, most customers were bewildered by the invention that in time became as standard and familiar as fast food itself. The Snyders had to show customers how to use the drive-through; they even enlisted their young sons to help.

Astonishingly, Harry Snyder and In-N-Out have received scant recognition for coming up with what is essentially the formula for today's drive-through system. Certainly, others have come forward to claim credit. Dave Thomas, the founder of Wendy's, boasted that he invented "the first modern day, drive-through window," when he rolled out his second Wendy's location in Columbus, Ohio—but that was in 1971, a good twenty-three years after Snyder's invention. Even McDonald's first drive-through window didn't appear until 1975, when it opened its initial model in Sierra Vista, Arizona, near the Fort Huachuca military base. In 1951, Jack in the Box introduced its own intercom "food machine" in San Diego. Back in 1931, the Pig Stand had devised a kind of primitive drive-through where motorists drove in and placed their order with a young male order taker and then exited—but their system relied upon an entirely human ordering and delivery process.

A clutch of what were called "drive-up windows" didn't really appear on the scene with any kind of critical mass until the mid-1950s. Soon resourceful drive-in owners began deploying a host of electronic ordering devices that promised both novelty and speed. However, they were hardly streamlined or elegant, and were made with such items as vacuum tubes, bulky switches, and carbon microphones. In many cases, carhops were still used to deliver the food. One of the best-known of the time was Sonic America's Drive-in. In 1954, the Stillwater, Oklahoma, shop introduced its own electronic ordering service, billed as "Service with the Speed of Sound!" Motorists pulled up to a row of parking positions and ordered their food through a handheld line connected to Sonic's kitchen.

Soon, a host of electronic ordering devices with names like Aut-O-Hop, Dine-A-Mike, and Teletray came onto the market. Despite their proliferation, they did not come into play until several years after Snyder pioneered In-N-Out Burger's own two-way speakerphone.

As Esther Snyder once proudly told the *Los Angeles Times*, In-N-Out was known as "the granddaddy of the drive-throughs."

At the very least, Snyder's stamp on the car and fast-food cultures certainly popularized the device and served as a prototype for all other fast-food establishments. And while In-N-Out modestly bills itself as "California's first drive-through," in all probability it was the country's first as well. In-N-Out's "granddaddy" gave rise to an entire world of drive-through banks, liquor stores, pharmacies, restaurants, dry cleaners, and even drive-through wedding chapels. In fact, the drive-through became so pervasive that decades later, in the 1990s, car manufacturers began to outfit front seats with cup holders to facilitate dashboard dining. Fast-food restaurants created meals tailor-made for one-handed steering and one-handed eating free from spills and dripping. More than a half century after Harry Snyder premiered his invention, the fast-food industry spawned a host of high tech gadgets all founded on the same premise as that first two-way speakerphone. However, according to government records, Harry Snyder never took out a patent on his invention.

At the start of In-N-Out Burger, it was just Esther and Harry. The Snyders did everything themselves. They prepared the hamburger patties (using an ice cream scooper to mold them and then their own palms to flatten them), peeled the potatoes, sliced the onions and tomatoes, and stirred up the secret sauce (that Harry concocted and would spend years perfecting). Harry ran the gas grill; Esther was in charge of the books. She had a real talent for numbers and a mind like a steel trap. She used her home kitchen table as an office. The couple worked punishing hours, often logging fourteen- or fifteen-hour days. The long hours and hard work never daunted Esther, in large part because she never for an instant doubted her husband. "Anything he decided to do usually turned out well," she once said, "because he would work quite hard. He was a person who, if you gave him a job and it was difficult, he would figure it out and not let go until he knew it well."

On In-N-Out's first day of business, the couple sold a total of forty-seven hamburgers. During their first month, they sold two thousand, bringing in an estimated $1,100. In recalling those early weeks, Esther Snyder once said "Many cold, smoggy nights were spent during the first few months of operation, but it was worthwhile." During the early years, the Snyders made great personal sacrifices. They rarely spent money on themselves; in fact, it was some time before they bought a television. Instead, the couple funneled every cent they made back into the business.

From the start, Baldwin Park's locals lined up to eat at In-N-Out. Some of the Snyders' biggest fans lived just across the street at the trailer park on Garvey. A few of the Barnes Circus families still lived there and often a few could be seen rigging a high wire three feet off the ground to practice their act while others rode their unicycles around the park's perimeter. They'd have barbecues and picnics and run across the street to pick up bagfuls of In-N-Out burgers. In fact, a whole generation of kids who lived in Baldwin Park's trailer parks grew up on In-N-Out. "Our early support came from the kind people of Baldwin Park," Esther proudly declared. "Those were the days we made our many friendships with the residents of Baldwin Park."

Margaret Howard was sixteen years old when In-N-Out first opened up. Howard lived with her family on a chicken ranch in Baldwin Park. At seventy-five years old she still vividly recalled the first time she ate there. "I remember because I was in high school and they were real good hamburgers," she explained. "My father didn't like us to eat out. He wanted my brother and I to eat our meals at home, so we conned our friends to take us there." When that didn't work, Howard said that she purposefully missed the school bus from Covina High School and walked the ten miles home. "That way I could go to In-N-Out on the way back. I got chewed out for it when I got home," she laughed. "But I got to eat at In-N-Out." Soon enough, going to great lengths in order to eat one of their fresh and juicy burgers had become a common practice among In-N-Out's rabid fans.

At the same time, In-N-Out caught on with hungry motorists. During the evenings, In-N-Out benefited from the local produce growers who drove at night and stopped by the little Baldwin Park stand on their way to the Los Angeles Wholesale Market. Russell Blewett, a longtime Snyder family friend who later became one of Baldwin Park's mayors, put it this way: "That place was a gold mine from day one."

It was during those first, early days of operation that Harry created the formula that emerged as the standard for running In-N-Out. It informed the company's identity and was rigidly adhered to over the coming decades. It was not based on some fancy management methodology—rather, it grew out of Harry's own instincts and exacting personality. The system was based on three simple words: "Quality, Cleanliness, and Service."

Harry was a micromanager before the term existed. A rigorous taskmaster, he was not inclined to leave even the smallest details to others. From the start, he kept scrupulous records, tracking how many burgers were sold daily and noting how many paper cups were dispensed.

Harry was known to be fanatical about quality. He insisted on inspecting everything, and everything had to be done to his specifications. When it came to In-N-Out's beef purchases, Harry constantly visited his meat supplier where he'd watch the butchers cut up the beef and make sure he got exactly what he paid for. He treated his suppliers well and never tried to exploit his relationships. Deals were struck on a handshake and lasted decades, often ending only if the supplier went out of business—or failed to meet Harry's exacting standards. It was a policy that lasted for years. When it was discovered that a vendor had hidden a batch of substandard onions within a truckload of good ones, the supplier was unceremoniously dumped. Esther (who handled the invoices) always paid their purveyors on time. If a business partner came to her to collect his money personally, she wrote him a check on the spot.

A savvy businessman, Harry established Snyder Distributing, a small paper goods wholesaler that sold such items as paper plates and napkins. At the time, In-N-Out Burger was too small to benefit from economies of scale by buying such items in bulk like many of its larger competitors. Snyder Distributing gave In-N-Out a way to buy supplies at a discount while turning a profit as a wholesaler selling to other businesses.

A frugal and practical man in most respects, Harry was profligate when it came to purchasing the freshest, highest grade of meat, potatoes, and produce; he refused to sacrifice quality for the sake of profits. From the start, he was adamant about using only four to five slices of the thick, middle part of big, plump beefsteak tomatoes and onions. He demonstrated the same resolve when it came to using only the crisp inner leaves of the head of lettuce, throwing the rest away. It was a practice that never changed. "Mr. Snyder stressed quality from the first day he opened for business," his wife once observed. "No matter what the price, he believed that the customer deserved the best product he could produce."

When it came to cleanliness, Harry's zealousness perhaps only matched his fervor regarding quality. He didn't feel that it was beneath him to scrub the floor or pick up trash. He even insisted that the gravel drive-through lane be swept between busy times. In fact, one of the Snyders' innovations was the open kitchen. Behind a large glass window, In-N-Out's customers could see how the burgers and fries were prepared and cooked and kept scrupulously clean. Harry made sure that his workers frequently washed their hands, especially after each time they took out the trash. "That place was immaculate," remembered Lorraine O'Brien, who along with her husband later owned and lived at the Baldy View trailer park on Garvey. "I'd walk past that kitchen and never see a spot. And I never saw a dirty-necked boy working there in my life."

Harry didn't like sloppy burgers, either. There was a system for building the burgers; the secret sauce was spread generously on the bottom slice of the bun to prevent it from running off at the ends. That way it wouldn't drip through the paper wrapping when customers went to eat it. When it came to grilling the burgers, it was one minute on the first side and two minutes on the other so as not to lose the juices. When cooking the fries, the fryer was kept clean and floating pieces removed quickly. Salt was to be shaken while holding the container at shoulder length to ensure evenness. Buns were lightly toasted before the meat and onion were added, and each burger received two slices of tomato. Only those tomatoes that fit five wide in a specially designed box were deemed the right size for In-N-Out.

Not surprisingly, Harry did not believe in cutting corners. It was his belief that the main reason people came to In-N-Out was for the burgers—if the burgers weren't done properly, the customers wouldn't come back. From the start, In-N-Out ran a customer-driven shop.

While the Snyders were experimenting with selling burgers over a two-way speaker and drive-through lane in Baldwin Park, another couple—a pair of brothers, Richard and Maurice (Dick and Mac) McDonald—were testing out a new kind of eating place over in San Bernardino, forty-five miles east of Baldwin Park. Originally from New Hampshire, the brothers had moved to Southern California in 1930, and they worked various odd jobs before trying their hands as movie producers. Unsuccessful in Hollywood, they opened a small movie theater that quickly went under. In 1937, they opened a hot dog stand near the Santa Anita racetrack. Three years later, they opened their first drive-in restaurant with a $5,000 loan from the Bank of America. It was a typical drive-in with indoor seating, carhops outside, and a twenty-five-item menu. "Sometimes I like to play a hunch," said S. P. Bagley, the bank's manager. "And I have a hunch that McDonald's is going to make it big."

Their drive-in was in fact not a huge success until 1948, when the McDonald brothers noticed that postwar customers were growing disenchanted with carhops. They devised a new streamlined approach to fast food and called it McDonald's Famous Hamburgers. The McDonalds turned their kitchen into a mechanized assembly line; they got rid of their carhops and the indoor seating. The two service windows where those carhops once filled their orders were turned into service windows where customers placed their orders directly. The three-foot cast-iron grill was replaced with two stainless-steel six-footers that were both easier to clean and more efficient at retaining heat. Instead of plates and silverware (costly to replace and clean), they used paper bags, wrappers, and cups. And they trimmed the menu to nine items, featuring hamburgers and cheeseburgers, three soft drinks, milk, coffee, potato chips, and pie.

While the Snyders remained focused on their burgers and keeping their customers happy, the McDonald brothers concentrated on keeping costs down and volume high. They introduced a slew of innovations intended to simplify and speed up the process of preparing and churning out burgers. For instance, they built a machine that could make their hamburger patties based on a device originally designed to produce peppermint patties. "Our whole concept was based on speed, lower prices, and volume," Dick McDonald once explained. "We were going after big, big volumes by lowering prices and by having the customer serve himself."

Inspired by the cost-effectiveness and success of McDonald's, other fast-food places began springing up. Soon a number of entrepreneurs launched their own fast-food restaurants based on the self-service style of McDonald's. This new way of eating quickly spread to the rest of the country—the era of the fast-food restaurant began in earnest. Family-owned and -run burger shacks were springing up everywhere, planted like trees along the growing off-ramps of California's equally expansive highway system.

Soon establishments with names like Wendy's, Burger King, Taco Bell, and Kentucky Fried Chicken were in business, blanketing the country, moving east from the undisputed fast-food capital of Los Angeles. As they flourished, they were transformed from homey, family-run establishments to regional, national, and eventually international chains dedicated to coming up with ever more efficient ways of cooking and serving massive quantities of food.

Everywhere, that is, except at the little In-N-Out Burger stand on Garvey Avenue where Harry Snyder had made a promise to himself that he had no intention of breaking: "Keep it real simple," he always said. "Do one thing and do it the best you can."

CHAPTER 4

In 1951, Allen Teagle was a teenager itching to join the navy when Harry Snyder hired the restless young man to work part-time for him at the Baldwin Park In-N-Out Burger. By then the Korean War was in full swing, but Teagle, who was just sixteen years old at the time, had to wait several months until his seventeenth birthday had passed before he could formally enlist. "Harry offered me a regular job," he explained. "But I told him that I was going in the navy and that I was just waiting until I was old enough. He was all right with that. He said, 'We can use you for a while.'" Snyder's casual reaction was something of an understatement.

In just three years, In-N-Out Burger had become an unqualified success. Long lines of cars snaked along the gravel drive-through lane at all hours, frequently causing "burger jams." It was a hometown hit that attracted customers from neighboring suburbs. Motorists on their way from one point to another often ended up at the little burger joint and not only made a point of returning the next time around but also began telling friends and associates about the tasty hamburgers being cooked up in Baldwin Park.

The volume, the traffic, and the business generated by the boxy stand on Garvey Avenue was too much for Harry and Esther to handle by themselves, and the Snyders began hiring help. Harry seemed to be a shrewd judge of a person's character, and he employed a handful of local young men to work in four-hour shifts. He insisted that they be well-groomed and wear a uniform of clean white shirts and aprons, capped off with paper cadet hats. Harry had high standards, and while he was bighearted he also expected his hires to toe the line—his line. "He wouldn't put up with any foolishness or clowning around," Teagle recalled. "And Harry always made sure that everyone was polite to the customers."

When it came to hiring female employees, however, Harry put his foot down. His objection came down to two words: "monkey business." There was too much opportunity for grab ass and messing around, he believed. "No 'foolishness,' he called it," said Teagle. "Things were different back then." Clearly, Harry's theory predated workplace guidelines on sexual harassment. Aside from Esther Snyder, females did not work in In-N-Out stores for another thirty years.

Harry put Teagle to work right away, during the early evening shift from 6:00 p.m. to 10:00 p.m. "On a big night we'd sell two hundred burgers!" he recalled. "My first chore was to clean up the mess. The first thing I did when I arrived and the last thing before I left work was to pick up the trash and rake the gravel. The kids would come in their hot rods and there would be dirt gravel everywhere and ruts in the driveway. Harry wanted it smooth so that it would look clean and neat." When he wasn't out raking the gravel drive, Teagle could be found in a small room in the back where all new hires were trained. There he sliced tomatoes, peeled potatoes, and washed utensils and pans. "I did anything Harry needed done," he said.

In the early 1950s, when television ownership was still something of a luxury and programming limited to a mere few hours a day, Baldwin Park itself offered little in the way of hometown amusements. There was the Vias Turkey Ranch off of Frazier Avenue, famous for its huge commercial ranch and outdoor aviary, as well as the horse stables located across the bridge of the San Gabriel River that offered pony rides for a dollar an hour. But for the most part, the young people of Baldwin Park were left with long stretches of time on their hands; the tiny burger stand on Garvey quickly established itself as a local hangout.

Open until 1:00 a.m., In-N-Out Burger became ground zero for Baldwin Park's restive teens. There the kids parked, played their radios, sang, and danced. But Harry had no patience for teenage angst in his parking lot. "If the kids got rowdy, Harry would make them behave or ask them to leave," recalled Teagle. "If they got out of line or had too much to drink he ran them off. He was all law and order." To prove his point, once, after the kids had gotten out of hand, Harry kicked everybody out and put a chain around the property and locked it up. The teens were allowed to return only after they promised Harry they would behave, which they did.

As it happened, In-N-Out Burger came of age just as the new youth and car culture emerged. In Baldwin Park, the two met at the little stand on Garvey Avenue. During those prosperous postwar years, Detroit had produced a whole new generation of vehicles. With a glut of new model automobiles traveling down America's roads, there was a huge surplus of the older cars just lying around. These abandoned autos offered the perfect occupation for the legions of Southern California's car-obsessed youths who enthusiastically took their parents' beat-up and outmoded Model T Fords, Chevrolets, Hudsons, and other vintage-tin bodies and recycled them. Inventive and daring, they removed flathead engines, stripped door handles, eliminated transmission casings, appropriated spare fenders, and repainted the vehicles in bold colors and designs, giving birth to the American hot rod. Once built, the only thing left for the kids to do was to show off their four-wheeled metal peacocks.

All across Los Angeles, hot-rodders and car enthusiasts converged on the plentiful neon drive-in burger palaces with their brightly lit interiors, wild angles, and giant V-shaped car canopies. During its ten-year heyday until 1968, the famous Harvey's Broiler (later renamed Johnie's Broiler) in Downey regularly attracted a parade of five thousand cars on weekend nights. "It would take thirty minutes to get to Firestone Boulevard and another twenty minutes just to get through the parking lot," remembered Analisa Hungerford, a Long Beach community college teacher who in 2007 helped spearhead a movement to save what remained of the site from demolition. "That place was wow. It was the first place that people snuck out of their houses to go to; there were so many first dates that happened there. It was just the jewel of Downey."

Showcasing hot rods on a Saturday night soon gave way to drag racing. Much to the chagrin of the authorities, teens and young men in their inventively modified and souped-up dragsters hit the boulevards on the clear, dry evenings and weekends with but a single purpose: to own a quarter-mile of asphalt in the shortest time possible. In-N-Out Burger became a crucial stop on the colorful but illegal street racing circuit that materialized after the war. It didn't take much for a race to begin. Usually two drivers met up and one popped the question to the other: "Do you want to drag?" That was it. Often, the losers lost more than just the race—many lost their cars, too.

Building high-powered, mean machines that were chopped, flamed, and louvered evolved rapidly. It was a world that author Tom Wolfe brought to mainstream attention in his 1964 article for *Esquire* magazine titled "The Kandy-Kolored Tangerine-Flake Streamline Baby." By then, of course, hot-rodding had caught on in the rest of the country. In Southern California, there seemed to be no shortage of cars or parts or for that matter young men with the desire to create their own hot rods.

One of them, the legendary hot rod builder Pete Chapouris, got his start cruising the boulevards of the San Gabriel Valley with his friends. Chapouris, who would go on to establish the celebrated custom garage Pete & Jake's Hot Rod Parts in Temple City (where he customized a 1934 Ford three-window coupe into the iconic cruiser featured in the movie *The California Kid*), once recalled those early intoxicating days by mapping them out in a single sentence. "We'd start at the In-N-Out on Valley, go straight west to Farmer Boys, out on Colorado to Bob's in Glendale before turning around and going east to Henry's in Arcadia."

Around the same time that dragsters were meeting up at In-N-Out Burger and spreading the word, the Baldwin Park burger joint began catering to another quintessentially Southern California phenomenon: surfers. Growing rapidly in popularity during the 1950s, surfing had by the 1960s exploded into a full-blown cult with a language, clothing, music, and a lifestyle all its own. In an almost religious ritual, surfers woke before dawn, strapped their waxed boards onto their cars and vans, and headed toward such fabled haunts as Malibu, Redondo, and Huntington beaches where they rode the swells of the Pacific Ocean. Many lived inland in the Valleys, and so commuting an hour to catch a wave before the sun was up was fairly common. Afterward, a group of hungry surfers packed up their cars and headed for In-N-Out. Soon enough, word spread among the hang-ten crowd, and ending up at an In-N-Out Burger stand became part of the surfing experience.

In those early days, In-N-Out Burger also earned a following among Hollywood glitterati who happened to discover the stand with the red and white awning while traveling between Los Angeles and Palm Springs. During the 1950s, well-heeled and big-name celebrities like Frank Sinatra, Kirk Douglas, Lucille Ball, and Liberace liked to frolic at the desert resort that became known as the "Playground of the Stars." Many had second homes in Palm Springs. On their way to the desert or back to the city, stopping at In-N-Out to grab a burger became part of the routine.

During their brief marriage (presumably before Elizabeth Taylor entered the picture), Eddie Fisher and Debbie Reynolds were known to visit often on their way to the desert. Once, Esther Snyder recalled the time when the singer Dinah Shore and her husband, George Montgomery, a stuntman and actor in Westerns, dropped in on their little In-N-Out joint. "They were having car problems, so Dinah came over to the house to watch her own TV show," she said.

After a friend introduced Bob Hope to In-N-Out, the legendary comedian became a lifelong fan. "More than anything, my dad loved an In-N-Out Burger," his daughter Linda Hope recalled. "He called them 'in and outers,' and he had to have his fix several times a week. He always got the cheeseburger and french fries. He used to say, 'They're just so fresh.'" In fact, Hope became such a devotee that at one point his daughter claimed he tried unsuccessfully to buy stock in the company. "He was really interested. He felt he was personally responsible for a number of their sales, and he just felt it was a great product."

When Hope turned ninety-two in 1995, his family hired an In-N-Out cookout trailer to cater his birthday party. "Dad continued to drive to one well into his mid-nineties," Linda admitted. And when Hope could no longer drive himself, she said, "I know he used to send people out there to pick it up for him and the staff." In fact, according to his daughter, Hope ate In-N-Out burgers regularly until he died on July 29, 2003, at the age of one hundred.

In-N-Out Burger was busy, popular, and cool. Despite the extra help, Harry's fastidious nature did not allow him to sit back and relax. Harry Snyder was the first person at the shop in the morning opening the store, inspecting the sacks of potatoes for the fries, ensuring that everything ran smoothly and to his exacting specifications. It wasn't unusual to see Harry picking up trash by the side of the road or leafing lettuce. And he was the last person there at night. When he took a break, he'd go in the tiny room in the back of the store and play checkers with friends.

Since the Snyders' house was directly across the street from In-N-Out, he rarely seemed to be able to relax and call it a night. In the evenings, Harry regularly kept an eye on the restaurant through his living room window. Frequently, he'd be sitting on the sofa watching TV when all of sudden he would stand up and sprint over to the shop and pitch in. As Allen Teagle tells it, "He was just a really hard-working guy. And he was always there when I was there."

From the very beginning, Harry offered his young hires more than just a job. For starters, he paid them well. When In-N-Out first started, California's minimum wage was sixty-five cents an hour, but Harry paid a dollar an hour, plus one free hamburger per shift. He believed in paying for quality, and that included wages. As Esther later explained, "They take your orders and make your food. They're so important, so you want to have happy, shining faces working there."

Early on, Harry offered his workers the opportunity to build a career. If he sensed promise, he did what he could to help the young men develop. He always tried to give his workers reasonable goals to strive for, promoting them when the goals were met. While everyone had to start at the bottom—cleaning up trash and peeling potatoes—everyone was also given the opportunity to advance. A number of teens who came to work part-time for the Snyders ended up making their careers at In-N-Out.

One of them was Chuck Papez. He got his start in the Baldwin Park kitchen as a teenager in 1954 and worked his way up to store manager. By 2000, he was one of the company's corporate managers. So satisfied was he with his life at In-N-Out that Papez later encouraged his brother, his two children, and a nephew to join the company as well. "It was special right from the beginning," he explained forty-six years after he first began peeling potatoes at In-N-Out. "It was family."

Harry was a man of his word; hard work was always rewarded. The meritocracy that he established in those early days hardened into standard operating procedure in the years to come. While his wife, Esther, hewed closely to a strong, unwavering faith in God, it might be said that Harry's faith was in hard work.

Harry had little interest in or patience for trends or fads, and he instituted a series of management practices that would not only contribute greatly to In-N-Out's future success but that were, in many respects, ahead of their time. Years before business schools discussed managerial terms like customer-relations management, worker empowerment, or profit sharing, Harry Snyder put these and other concepts into practice. He gave his associates a measure of ownership in the enterprise and he remunerated them handsomely for their ability to meet their targets and surpass them. Harry put a huge emphasis on customer satisfaction. In-N-Out workers were instructed to always smile, look their customers in the eye, and maintain a level of professional courtesy with every guest. Long before Starbucks, the Snyders called their customers "guests." Harry rationalized that if your customers don't come in or come back, there is no business, period. Harry drilled into his workers the singular importance of quality and simplicity. His maxim, "Do one thing and do it well," was not only repeated with some frequency—it was strictly adhered to. Harry kept the menu simple and streamlined. He didn't see much advantage in introducing new items or tinkering with the ones they already offered. Harry Snyder had picked up the rhythm of human interaction, and his business philosophy was based on it. If you treated people fairly and rewarded them accordingly, he held, they would do likewise.

In many ways, In-N-Out Burger was an employee-driven company. The Snyders displayed an uncommon respect for their workers. The couple never viewed their business as just some little burger joint—in much the same way, they never looked at their workers as just employees but saw them as part of their own growing, extended family. The Snyders made sure to know each individual by name. In fact, they banished the words "employees" and "workers" altogether and instead referred to them strictly as "associates." The result was that from the outset, In-N-Out had the feel not of a workplace but of a joint enterprise in which everyone shared.

As exacting a fellow as Harry was, his diligence was balanced by an equal sense of affectionate generosity. From the very beginning, Harry and Esther believed that their associates were integral to the success of their business, and they also believed in sharing that success with them. During In-N-Out's early days, the Snyders doled out ten-dollar bonuses to their associates. "My husband always believed that employees should share in the profits," Esther later explained. "He believed that if you performed well on the job, you were worth the extra money."

As In-N-Out grew, so did the Snyders' largesse. Every Christmas, Harry walked into the Baldwin Park branch of the Bank of America carrying a list of savings bonds for the teller to type up: one-hundred-dollar bonds for managers; twenty-five dollars for their spouses and children; and twenty-five-dollar bonds for all other associates. Harry was unusually generous in extending a hand to numerous employees; he helped many of them get loans for houses and cars, sometimes lending them the money himself. "It just depended on how long he knew you or how long you had been there," explained Bob Meserve. "He helped everybody."

Throughout the year, the Snyders threw numerous gatherings, picnics, barbecues, and Christmas parties in their own home for their associates and their families. The events were always warm and inviting, and as the years wore, the gatherings became bigger and more elaborate—but they managed to retain the feel of a large family gathering. The Snyders always treated their associates with respect, giving them a sense of ownership. In return, In-N-Out's associates felt incredibly loyal to the Snyders. It is likely that the most important decision that Harry and Esther Snyder made was the loyalty that they built between In-N-Out and its associates.

CHAPTER 5

It wasn't the Snyders' desire to build an empire, to increase their revenues, or even to show up their competitors that propelled the couple to expand—rather, it was their associates. The Snyders had created a web of dedicated (and well-paid) employees trapped inside a one-store wonder with nowhere to go. Part-time workers became full-time workers and the full-time workers started to become managers. Initially, the Snyders assumed that most of their associates would gain experience at In-N-Out and then move on to something else or even to another fast-food restaurant. Instead, a bottleneck of talented, qualified people emerged who—instead of moving on—wanted to move up within In-N-Out. It became clear that the person standing on the third rung of the ladder could not reach the second and the person holding onto the number two spot was never going to climb to the top unless somebody above either quit, which seemed ever more unlikely a scenario as time went on—or the Snyders got rid of someone.

Those who were close to Harry and Esther explained that it was never the Snyders' intention to build a large chain of In-N-Out Burgers. According to Bob Meserve, his uncle was a fairly straightforward and conservative man. "Harry's thinking was that he was just trying to make a living for his family. He wasn't interested in being a rich man. The only reason he built the stores was because he paid his people well and they didn't want to leave. He wanted them to save their money and go into business for themselves."

A stickler for quality, Harry had implemented a system in which everything was fresh and made daily. He knew everyone who worked for him—and each and every one of his suppliers—on a first name basis. The system that he laid down was militarily focused. If In-N-Out expanded, it was vulnerable to dilution of the quality standards he so vigilantly held. Despite the obvious financial advantage inherent in expanding into a chain, there was also the very real possibility that in doing so Harry might have to give up some measure of control. This was the dilemma set before the couple: stay the course as a one-shot mom-and-pop shop or grow into a larger network of In-N-Out Burgers.

In the end, it was the Snyders' commitment to looking after their people that prompted In-N-Out Burger's growth into a chain. In those early years, the couple opened new stores as a way to reward the hard-earned dedication of their longtime associates who wanted to remain within the In-N-Out fold. A significant number of those early associates who began in In-N-Out's tiny kitchen as potato peelers ended up staying on for decades. Looking back more than fifty years after he turned in his In-N-Out apron to join the navy, Allen Teagle said wistfully, "If I hadn't joined up, I would have stayed there a lot longer."

The In-N-Out expansion began slowly in 1951, three years after the Snyders opened for business in Baldwin Park, and through the years the rollout of new stores continued at a glacial pace. The second In-N-Out Burger opened in Covina, a small town about four and a half miles northwest of Baldwin Park that bordered Irwindale to the west, Azusa and Glendora to the north, and San Dimas to the east. By 1952, the Snyders had opened an additional three shops in La Verne, West Covina, and Pasadena. The expansion resembled an outstretched hand across the San Gabriel Valley, with Baldwin Park resting in the middle of the palm.

As the chain grew, the business relationship between Harry and his partner Charles Noddin soured. As is typical among business partners, the two men had very different ideas about the future direction of In-N-Out as it expanded. Harry insisted on maintaining a quality product at a reasonable price, while Noddin reportedly wanted to increase prices and cut costs. At the time, Noddin's strategy was becoming common practice among fast-food chains—but Harry was determined not to go the way of every other burger and hot dog stand. He believed that his way was the right way. A stubbornly independent sort, Harry had a real aversion to ceding to the opinion of others, especially when he thought he was right.

By 1952, Harry and Charles Noddin dissolved their partnership for good and split up the existing stores. Harry kept the In-N-Out Burger name and three of the stores including Number One in Baldwin Park. Noddin went on to open another burger chain in Pasadena. But the experience taught Harry a lesson; he insisted that In-N-Out remain forever independent. After that, as one friend said, "Harry swore he'd never sell out or take in a partner." Harry's nephew Bob Meserve put it this way: "He was happy with the golden goose."

Harry Snyder was extraordinarily discerning when it came to deciding on when and how to open a new store. Essentially, it came down to three decisive factors: his associates, a location, and the balance sheet. Regarding the first, the associates who were tapped to manage had initially gained sufficient experience working under Harry to carry out and follow to the letter his very specific operating procedures. Each had worked his way up. Everyone was trained in the little room in the back, where they were taught everything from the correct way to put on an apron to the best method of cutting onions with the hand slicer. And Harry didn't open a new shop until he had a manager who was ready to run it properly.

Secondly, with respect to locations, initially the stores were spaced out close to Baldwin Park so that Harry could maintain his standard of delivering fresh meat and produce on a daily or near-daily basis. It also allowed him to preserve his oversight of the operations. Mornings, afternoons, and evenings, Harry was known to drop by his In-N-Outs unannounced. He kept a close watch on his growing chain. Harry and Esther liked to check on the shops often; it gave them the opportunity to stay in close contact with all of their associates. The couple liked to know each of them on a first name basis. Visiting the stores, chatting with the associates, and checking to be sure that procedures were followed delighted Esther in particular.

In catering to the car-reliant customer, Harry focused on putting his drive-throughs in highly visible, heavily trafficked areas such as major intersections and thoroughfares. As time wore on, new stores were placed right next to off-ramps of the fast-expanding freeway system. In choosing the new sites, as in most things, Harry relied on his intuition—but luck played a part, too. The growing Southern California freeway network became a significant factor in In-N-Out's own rising popularity and reputation.

At one time, Los Angeles had an enviable large-scale trolley network. The Pacific Electric train, with fifteen hundred miles of track, stretched seventy-five miles from San Bernardino to San Fernando and south to Santa Ana. Before it was dismantled and relegated to history, during the late 1930s, the system was carrying some 80 million passengers a year. But following the war, the rail lines soon ground to a halt. Indeed, on October 14, 1950, at 11:23 p.m., the Pacific Electric, once the pride of Baldwin Park, pulled into the town's depot for the last time.

The demise of the fabled Red Line in Southern California is generally connected to a complex corporate plot hatched by a consortium that included General Motors, Standard Oil, Phillips Petroleum, and Firestone Tires, all of whom allegedly conspired to wipe out the rail networks and create a transportation system heavily dependent upon the automobile. The scheme as it was later uncovered began in the 1920s. GM and its associates created a front company, National City Lines, with the specific goal of purchasing streetcar lines in a number of large American cities including Detroit, New York, and Los Angeles—whereupon the lines were dismantled, shut down, and replaced with a bus system (the buses being manufactured by GM). It wasn't long before the population was left with essentially one choice: to drive themselves.

The scheme—rehashed over the years and labeled one of the great scandals of the postwar years (although others labeled the conspiracy myth)—also served as the plot of the movie *Who Framed Roger Rabbit*. The scandal shot straight to the Supreme Court, and GM and several of its cohorts were later indicted on federal antitrust charges. The presiding judge in the case fined GM and the other companies involved $5,000 each and the executives who hatched the plot were forced to pay a one-dollar fine.

In any event, the car ruled and transformed California. As the Great American Streetcar Scandal and its aftermath unfolded, a number of other dynamics contributed to the primacy of the automobile and the end of the trolley lines. For one, many of the lines began operating buses themselves, and just as many of them failed to modernize their maturing, atrophying equipment. Moreover, car ownership was steadily increasing. As the noted man of letters William Faulkner observed in 1948, in his novel *Intruder in the Dust*, "The American really loves nothing but his automobile—not his wife, his child, nor his country, nor even his bank account first." At the same time, tens of thousands of miles of new highways were being built across the country.

By the time President Dwight D. Eisenhower signed the Interstate Highway Act in 1956, the automobile had become a huge presence in American lives. The commander in chief concluded that "newer, multi-lane highways were essential to a strong national defense." Eisenhower had first become convinced of this as a young lieutenant colonel during the 1919 Transcontinental Convoy (in which eighty-one motorized army vehicles traveled across the country from Washington, D.C., to San Francisco in sixty-two days). His opinion was further galvanized during World II when he saw firsthand how fast Allied troops could travel across Germany's four-lane Autobahn.

The interstate act became the largest public works project in American history. The initial cost estimate was set at $25 billion to build over forty-six thousand miles of road in some twelve years; in actuality, it ended up costing $114 billion and took thirty-five years to complete. In In-N-Out territory, the San Gabriel Valley, a thirty-mile span of the Interstate Route 605 freeway, estimated at $60 million and to be completed by 1966, was planned to provide an unbroken artery that would run from the San Diego Freeway (405) in Orange County to the San Bernardino Freeway (10) in Baldwin Park. The new stretch of the I-605 would serve "territories as varied and fascinating as any in the State. Beach playground, poultry and dairy farm, oil field, bedroom suburb, college campus, country club, historical landmark." This was how the official journal of the California's Department of Public Works assessed the new eastern loop of metropolitan L.A.'s interstate.

Heralded as an economic miracle for the country, the highway system was expected to flatten the rural/urban divide. In reality, it accelerated the suburbanization of America. For businesses, the freeways dissecting the land literally transformed the landscape. The American chain stores already on their way to redefining the country were certainly aided by the new interstates. The asphalt tributaries developing all over the San Gabriel Valley were a boon for the fledgling In-N-Out chain.

When in 1954 the I-10 freeway cut through Baldwin Park, forcing the Snyders to tear down their stand and rebuild a new one a short distance from the original (it also left Baldy View with only forty-seven trailer home slots), Harry came up with another innovation that rivaled as well as complemented the two-way speakerphone; the new shop was designed with a double drive-through. The rebuilt In-N-Out sported two driveways flanking the open kitchen and four two-way speakerphones. The new format helped to speed the long lines that had been creeping along at a snail's pace as a result of the constant volume. As it turned out, a number of fast-food operators ended up adopting the double drive-through layout as well. For one, it was cheaper than the standard fast-food format to build, providing higher profits with lower overhead and requiring considerably less square footage than the average fast-food restaurant.

Actually, the suburban network of In-N-Outs that followed alongside the new highways also happened to dovetail nicely with Harry Snyder's third criterion for placing his stores; he abhorred debt. The new shops (in addition to their general proximity to Baldwin Park) were all situated in peripheral, suburban areas. They were not (as many other early fast-food places began) located in downtown centers next to factories or offices. In-N-Outs sprung up in the growing rural, postwar bedroom communities of the San Gabriel Valley. The suburban sites were cheaper than urban ones, especially as Southern California real estate values soared. Harry insisted on using cash, not credit, to open each new restaurant. Harry followed the old rules. He built one store, saved money; he built a second store and saved more money. He didn't open another until he could afford to and had the trained managers to run it—that was the Harry Snyder way. He didn't take out a loan. He didn't take on debt. He was beholden to no one.

While Harry's desire to own the land underneath his restaurants certainly limited the pace of In-N-Out's rollout, it also proved a remarkably shrewd financial move. In the 1960s, when Harry had expanded the successful chain to seven units, a former colleague congratulated him on his achievement and Harry responded by saying, "By God, they're all bought and paid for, too."

In-N-Out's growth strategy—actually its entire strategic approach to business—offered an interesting counterpoint to the industry that was evolving around it. Over in San Bernardino, the McDonald brothers' streamlined and automated drive-in at Fourteenth and E streets was generating a huge volume of customers and word spread quickly. After *American Restaurant Magazine* ran a cover story on McDonald's phenomenal success in its July 1952 issue, the brothers were inundated with inquiries from restaurant owners asking to visit and eyeball the operation themselves. Almost immediately, offers poured in to copy or buy the McDonalds' methods outright. Soon enough, the brothers began licensing their Speedee Service System.

They took out a full-page ad in a trade magazine that declared: "This may be the most important 60 seconds of your life."

Would-be burger moguls descended upon San Bernardino from all over the country. Within two years the brothers had haphazardly sold fifteen franchises—the first to a gas station retailer in Phoenix. Unofficially, the McDonalds' celebrated Speedee System was widely duplicated, and copycat versions began cropping up all over the country. The success of McDonald's spurred another San Bernardino resident and former World War II marine named Glen W. Bell Jr. to turn his trio of Mexican food stands (called Taco Tia's) into what eventually became Taco Bell. In 1952, duly impressed with McDonald's operations after a visit, Matthew Burns insisted that his stepfather, Keith G. Cramer (the owner of Keith's Drive-In Restaurant in Daytona Beach), fly out to see it for himself. The Insta Company's automatic broiler and milkshake mixer cinched it for Cramer. The pair returned to Florida (with rights to both machines) and a year later opened up their own self-service drive-in called Insta-Burger-King in Jacksonville featuring flame-broiled burgers—and immediately began selling franchising rights. Five years later, the chain was renamed Burger King. It was the same year that David R. Edgerton Jr., a Dade County franchiser, and his partner James McLamore (who took Burger King national) introduced "The Whopper."

Then in 1954, a fifty-two-year-old former paper cup salesman from Oak Park, Illinois, named Ray Kroc decided to make a trip out to San Bernardino. Kroc, as the story goes, had been selling a five-spindled milkshake maker called the Multimixer to neighborhood drugstores. His curiosity was piqued when the McDonald brothers ordered eight Multimixers including replacement and spares. That meant the little burger shop was making something on the order of forty milkshakes at a time—this as Kroc's bread-and-butter customers, the drugstores, were fast going out of business, having fallen victim to (among other factors) the growing success of the new fast-food drive-in. Upon Kroc's visit to the Downey McDonald's on the corner of Florence and Lakewood, he was transfixed. It was the third of the brothers' newly franchised sites. There was no indoor dining area, and the restaurant was fronted by a large walk-up service window, studded with shiny red and white tiles, topped with a jutting raked roof, and of course framed by a giant set of thirty-foot parabolic golden arches.

During the lunchtime rush, 150 cars crowded into the parking lot. When Kroc saw the speed of the operation and the volume of customers he declared, "Son of a bitch, these guys have got something. How about if I open some of these places?"

To Kroc's great good fortune, Dick and Mac McDonald were content with their business and not interested in taking on the kind of national business that Kroc envisioned. But that didn't mean that someone else couldn't expand the operation further. Kroc was a consummate salesman, and he soon convinced the McDonalds to make him their exclusive franchising agent to take the company national. On April 15, 1955, Ray Kroc opened his own McDonald's drive-in in Des Plaines, Illinois, and officially established the McDonald's Corporation. Over the next five years, Kroc built a chain of 228 McDonald's that were grossing $56 million a year. However, the relationship between Kroc and the McDonalds was not nearly as successful. Kroc was only earning a paltry 1.9 percent of the gross of all those McDonald hamburgers that he sold and 25 percent of that went to Dick and Mac McDonald. Soon the paper cup salesman had become tense and uneasy with the McDonald brothers. Finally, he had had enough of the relationship. In 1961, Kroc asked the brothers to name their price and persuaded them to sell him the company outright. Their terms: $2.7 million for the company and the name. Soon, Kroc not only boasted that he would open one hundred McDonald's each year, but he actually surpassed his initial bold claim.

Actually, it is interesting to note that as a result of his purchase of McDonald's national franchising rights, Ray Kroc was transformed into an American business icon largely credited with inventing a totally new form of dining almost overnight. However, according to a few industry pioneers, in fact it wasn't so much that Kroc franchised McDonald's but that the McDonald brothers were the first fast-food outfit to say yes to Ray Kroc.

On April 11, 1947, Alan Gamble, a former golf pro from Chicago, and his wife, Ellen, opened a hamburger and pie restaurant based on generations of original family recipes on a spot of farmland on West Pico Boulevard in Los Angeles. The Apple Pan made its home in a little cottage-style house with a centerpiece grill surrounded by a U-shaped counter encircled by twenty-six red vinyl stools. Back then, the area was surrounded by orchards and bean fields and looked nothing like the pricey Westside real estate it eventually became. Like the Snyders, the Gambles were fanatics about quality and they declared it with a neon sign at the front door: "Quality Forever." Soon enough, the Apple Pan earned a following for its savory hickory burgers, its tuna salad sandwiches wrapped in wax paper, and its brick-solid apple and pecan pies topped with thick whipped cream.

But according to Alan and Ellen's daughter, Martha Gamble, Ray Kroc had made a visit to their eatery in 1949—a good five years before his trip to McDonald's—and made them an offer. "He liked our concept and wanted to incorporate our ideas into a national franchise," she said. "My parents said no. They just wanted one place. They wanted to do something and do it well. They said you can't keep your eye on things like you can at one place." Gamble, who now runs the Los Angeles cult classic with her daughter Sunny Sherman (exactly as her parents had, right down to the original tartan wallpaper, paper cone soda pop cups, and staff who have been with the Apple Pan for over fifty years), recalled, "I do remember that mother and dad talked about it. But they never got into the business to get rich. The whole idea for them was to make the best hamburgers, sandwiches, and pies that they could make."

Even the Snyders' old friend Carl Karcher said that at one point Kroc had extended a similar offer to take on his small clutch of hot dog stands and barbecues and turn them into a national franchise. "Ray Kroc became a good friend of mine," Karcher laughed. "But I said no way." Apparently, he misjudged In-N-Out's potential. Around the time Kroc visited McDonald's, he approached the Snyders, hoping to sell them one of his Multimixers.

Certainly Kroc wasn't alone. In 1952, an elementary school dropout, onetime farmhand, insurance agent, and railway fireman named

Harlan Sanders opened his first Kentucky Fried Chicken restaurant near Salt Lake City, Utah. Within ten years, Sanders (now known as the Colonel) had become a national phenomenon and his restaurant, with six hundred franchises, was the largest chain in the country. In 1964, Sanders (who had begun opening outlets in Canada and England) sold his interest in Kentucky Fried Chicken for $2 million to a group of investors led by John Y. Brown Jr.—the future governor of Kentucky. Seven years later, in 1971, Heublein Inc. acquired the company for $285 million. At the time, there were over thirty-five hundred Kentucky Fried Chicken restaurants dotting the globe—but Sanders never saw a penny from the sale.

That was the fast-food game in the 1950s—start, grow, franchise, sell. The decade is widely considered the industry's boom time. The dominant chains of today began in the 1950s. While the original chains were wholly owned, franchising offered operators a way to grow with moderate financial risk. Owners could replicate a successful business and at the same time the franchisees had a stake in the continued success of the enterprise. Both sides could potentially profit by mitigating the risk; the franchisor through charging licensing and other fees and sharing in profits, and the franchisee by investing in a system that already had a proven track record. The numerous family-owned fast-food stands that had sprung up following the war were now in an arms race of their own. Any number of mom-and-pop outfits were cherry-picked and greatly enlarged through franchising into regional, national, and later global chains. Few holdouts remained in the game for long. Those who didn't franchise or sell out were soon swallowed up by bigger organizations or simply disappeared. Some, like Harlan Sanders and Dick and Mac McDonald, sold out but realized only a small portion of the vast profits that the chains that they started eventually made.

In-N-Out Burger could have easily been one of them if not for Harry and Esther Snyder's steadfast philosophy. The couple was unmoved when Ray Kroc more than made good on his promise to open one hundred new McDonald's stores a year. Actually, the Snyders remained remarkably unconcerned. Harry Snyder had an extraordinarily lucid vision for In-N-Out Burger, and Esther Snyder shared her husband's view. If anything, the couple had demonstrated time and again that they were as practical as they were astute. They had no interest in selling or franchising In-N-Out Burger. If you did that, Harry insisted, you'd lose control and focus. And they weren't interested in using their small and growing fortune to create an even bigger pot of gold if that meant taking in investors or selling to a larger company. Early on, Harry Snyder had added another tenet to his management stockpile. He saw no point in sacrificing quality for profits. In the words of one longtime friend, "It really never was about the money for them."

CHAPTER 6

On February 22, 1951, Esther gave birth to the couple's first child, a son they named Harry Guy Snyder. The following year, Richard Allen Snyder, who was born on July 13, 1952, joined his older brother (whom everybody called Guy). The Snyder brothers were just two of the millions of babies born following World War II. American maternity wards were filled with armies of swaddled newborns while baby carriages stood guard at the gates of family homes and parks across the country. On May 4, 1951, a *New York Post* columnist named Sylvia F. Porter coined the term for the phenomenon of infant births by which this generation would thereafter be known: baby boom. "Take the 3,548,000 babies born in 1950," wrote Porter. "Bundle them into a batch, bounce them all over the bountiful land that is America. What do you get? Boom. The biggest, boomiest boom ever known in history."

America's baby boom persisted in tandem with the flush economy of the fifties and the Snyders brothers' young lives looked like something straight out of *Life* magazine at the time. Theirs was a relatively ordinary if not picture-perfect 1950s childhood: orderly, languid, and cushioned by material well-being. Growing up at their parents' knees at the In-N-Out Burger, the pair not surprisingly adopted the American icons of hamburgers and cars as their own. The burger stand just across the street from their house was practically the brothers' second home, and for a time, the drive-through was their playground.

There is a photograph taken of the family in front of the Baldwin Park In-N-Out shop sometime in the 1950s. Harry is clearly the central figure, and he is standing straight with his back to a prized speedboat that is parked in the lot. He is looking dead ahead into the camera. Wearing a crisp, short-sleeved shirt, Harry has his pants creased sharply, and a handyman's keychain is dangling from his belt. Esther stands slightly behind Harry and is gazing up adoringly at her husband. She's wearing a simple, checked shirtwaist dress and her right hand is folded around a large thermos while her left hand is grasping a paper cup. The Snyders' young sons are sitting in the boat and are dressed alike as miniature yachtsmen; each is wearing a captain's cap. Guy is in the driver's seat and his hands grip the steering wheel playfully. At the same time, Rich is sitting back, facing the camera directly like his father. His right arm is crossed over his chest and his hand is balled into a loose fist pressed against his mouth. His is a pensive expression well beyond his years. Above the family and slightly in the distance is an early In-N-Out Burger billboard. The photograph is the essence of the fifties' middle-class American family, and its myopic calm could be mistaken as a still from one of the era's best-known sitcoms—*Leave It to Beaver*.

Guy and Rich were by turns both spirited and shy. Friends described Guy as something of a free spirit, while Rich, they said, always had a smile on his face. In the morning, Rich went outside, and if he saw sunshine, he came back and said, "Mom, it's going to be a happy day." The two shared an appetite for light-hearted fun; Rich in particular liked to play practical jokes. Once he took a tractor and shaved his name into a neighbor's field. Just a year apart, the two boys fought constantly. Theirs was a dynamic of kinship and rancor, and they were highly competitive with each other. They seemed to like nothing more than playing to win—neither liked to lose. "They had a love/hate relationship," was how their cousin Bob Meserve described the brothers. "They were always fighting, but if you stepped in between them they'd both turn on you."

Guy seemed to have inherited his father's temper and irascibility, while Rich was straightlaced from the start. Early on, Rich displayed some of his father's natural business acumen and his mother's gentle and sociable nature. When it came to their parents' attention, however, Rich overshadowed Guy. Rich was designated the good child who did what he was supposed to do, while mercurial Guy was given to whims, often chafing at the discipline and control that were required of him. Yet it was Rich who frequently stood up for his older brother to their hot-tempered father.

Guy came to be considered something of a rebel. Rich was viewed by many as the favored son, perhaps because his temperament and outlook were more in line with his parents' expectations. It was a perception that only grew deeper with time. As more than one family friend described the two, "Rich and Guy were like night and day."

In a relatively short period of time, Harry and Esther Snyder had something that neither of their parents had achieved in their own lifetimes: wealth and financial security. The new prosperity enabled the family to move to a larger house in San Dimas, which they did not long after the I–10 freeway was built practically at their Baldwin Park doorstep. Thirteen miles from Baldwin Park, San Dimas was situated closer to the foothills of the San Gabriel Mountains. It was a desirable suburb, and certainly a more affluent one than the blue-collar Baldwin Park. The Snyders' new two-story house sat on an expansive ranch surrounded by avocado and fruit orchards on the corner of West Covina and West Badillo boulevards, but it was solidly middle class. There was nothing flashy about their house. As one of the brothers' childhood friends remarked, "They had tons of money but they sure didn't flaunt it."

Despite their financial success, the Snyders remained a hardworking, down-to-earth couple with straightforward pursuits. Harry enjoyed boating, making home movies, and traveling with the family: car trips across the U.S. (stopping to photograph every roadside attraction) were one of his favorite things to do. The couple wanted to expose their sons to a variety of experiences, and so they took the boys overseas on trips to Europe, too. Esther considered herself a lifelong student and was a firm advocate of education; this was something she hoped to pass on to her children. During the family's travels, she made sure that the boys learned about the different landmarks and places they visited. Extremely well-read, Esther was surrounded by piles of magazines and books, and she was always recommending something to others. When she was finished, Esther gave her reading materials away—usually to In-N-Out associates or the local library.

Most of the time, the couple was on the same page—except when it came to the armed forces. Harry insisted that the army was the best branch, while Esther firmly corrected him. "No," she said, "the navy was best."

Esther's role at In-N-Out was largely behind the scenes. Although Harry often credited his wife's ideas—as Esther's sister Virginia Stannard put it, "He thought her advice was worthwhile"—she remained incredibly humble about her own contributions. Esther was in charge of accounting, taking care of the bills, and all of the paperwork. Esther was like June Cleaver with a small payroll department and two kids. "I doubt if she ever spent a lot of money on herself," said one of the Snyder's longtime friends.

Perhaps as a result of the couple's own hardscrabble childhoods— or perhaps in spite of them—Harry and Esther were determined that their boys would not grow up to be spoiled. As youngsters, Guy and Rich were given chores and put to work at In-N-Out doing odd jobs like picking up trash and later peeling potatoes and working the grill, gradually moving up the ranks—and they were paid just like everybody else. Early in life, they learned the value of hard work and the rewards that accompanied it.

Perhaps more than anything it was the Snyders' work ethic that was drilled into their sons. Once, when Rich was about eleven, he wrote "HELP CLEAN UP" on a chalkboard in the back room of In-N-Out. Harry misread the note. He thought his son was referring to the associates as help and gave him a bitter verbal lashing. "Never call anybody help," he hissed. "These people are our associates and we embrace them like family." Rich was eventually able to clear up the misunderstanding with his father, but it was a lesson he never forgot.

Harry Snyder's parenting had much in common with his management philosophy. He was a hard-driving perfectionist who lived by the very high standards he set and expected his sons to follow suit. Esther was usually ready to defer to Harry's authority (at least on the surface). When it came to her children, she was quite accommodating—even indulgent. While Harry's own father, Hendrick, was a tough disciplinarian who on occasion resorted to fisticuffs, Harry himself believed in teaching obedience and moral behavior to his own sons for the most part by imposing rigid expectations. "They respected their father," explained a longtime friend. "But truthfully, I think they were also a little afraid of him."

It may have been that Harry still held onto some of the values of his parents' generation. Although Dr. Benjamin Spock's best-seller *The Common Sense Book of Baby and Child Care* was changing the way that postwar parents raised their boomer children by encouraging natural affection over strict discipline, Harry wasn't generally given to blindly following the suggestions of others—especially when he had his own sense of the order of things. Indeed, Harry did what was not an uncommon practice among fathers who wanted to instill values and discipline in their sons; he sent them to military school. For a time, the Snyder boys attended Brown Military Academy in the neighboring San Gabriel Valley suburb of Glendora. An all-male Christian school, it was modeled after the U.S. Military Academy at West Point. At the time, it was the largest military school on the West Coast.

At the academy, the students were given full dress uniforms based on those worn in the War of 1812. During ceremonial occasions and the school's monthly parades, students were required to wear their tight gray coatees, trousers, cross belts, white gloves, and tar bucket shakos. In sun-splashed, 1960s Southern California, one can only imagine how that kind of anachronistic display went over with teenage boys whose major interests had turned to fun and fast cars.

Harry had the opportunity to find out at least once, when his sons played hooky from school. It was an infraction clearly not in line with the academy's stringent standards. However, for some reason the administration did not inform the Snyders of the transgression. According to an intimate, Harry eventually found out. A man of volcanic temperament, Harry Snyder appeared at the school immediately to set Brown's supervisors straight. "He reamed them," explained the friend. "He yelled, 'Don't ever cover for my kids! When they do something wrong, I want to know.'"

When Brown was sold to Azusa Pacific College and the campus closed in 1967, the Snyder boys were enrolled in Bonita High School, a public school in the San Gabriel Valley suburb of La Verne. At Bonita, where Guy began as a sophomore and Rich a freshman, neither of the brothers appeared to distinguish themselves academically. In the *Echoes*, the school's yearbook where their activities are cataloged, the only evidence of Guy can be found in his annual class pictures. Of the two, Rich was the more active. In 1967, while still a freshman, Rich was a member of the school's rocket club and the varsity football team. Guy Snyder, according to classmates, was the better-known brother among the students on campus, perhaps because of his outsized antics.

In line with his growing rebellious streak, Guy was known to ditch class. A thrill-seeker, one of his favorite after-school antics was doing donuts in a tractor in the large field in front of the family's house, just going around and around in circles and conducting mini-races that usually ended up in the street, incurring the wrath of the local sheriff. "He was fun-loving and ready to party," remembered Elaine Setterland, a classmate at Bonita. "Guy had this green muscle car, and after school all the kids would jump in and they'd go to the In-N-Out Burger." Although there were only a few stores in the chain at the time, already the burger shops had a large and devoted local following. At least twice a year during big school events, the Snyders brought in a trailer and served In-N-Out Burgers to all the kids—no doubt scoring points for the Snyder brothers with the Bonita High School student body.

The Snyder boys seemed to be surrounded by friends, including a group who lived at the McKinley Home for Boys (an orphanage located not far from the Snyders' home) whom Rich and Guy had befriended. One of them, Wilbur Stites, became the Snyders' foster child after Rich asked his parents if the twelve-year-old boy could come live with them. Although Harry and Esther raised him like a third son, Wilbur always called them Mr. and Mrs. Snyder. He was a bright boy, and the Snyders offered to pay his college tuition, but Wilbur declined, preferring to work in In-N-Out's maintenance department.

According to Susie Ericson (*née* Nissen), one of Rich's Bonita classmates, the brothers were "pranksters and fun-loving and they had a lot of friends." But they were also, she said, "terribly shy and awkward." And they were greatly overweight. In his football photograph published in the Bonita High School yearbook, Rich, although just a freshman, is clearly the largest person on the team.

Ericson said that Harry Snyder was fairly strict about allowing his children's friends over to the house. Like most things, when it came to his sons, Harry was no-nonsense through and through. He laid down the law when it came to sweets, banning them from the Snyder home. At night, he'd wait up for Guy and Rich, punishing them if they didn't make their curfew. Esther was more lenient; she welcomed everyone, feeding them, cooking up steaks, and not telling Harry when the boys came home late.

The Snyder boys spent a great deal of time over at the Nissen house on Oakmead Lane in La Verne. The Nissens were a large, loving and boisterous family—in addition to Susie, the Nissen siblings included Kenny, Johnny, and Kathy. In particular, Guy gravitated toward Kenny. The boys were the same year in school and shared a love of cars. "They were always out there in the driveway," remembered Kathy. "They were just working on cars and fiddling on cars." As Susie explained, "Guy always loved having people around him. He was lonely and overweight and he didn't feel confident. When he found a true friend, he was really good to them."

During their high school years, the brothers possessed one-half of the all-important teenage social equation: a set of wheels. However, when it came to the other half—girls—Guy and Rich came up short. Painfully timid around the opposite sex, neither dated much or had girlfriends that anybody could recall. They were, however, at ease with Susie and Kathy Nissen, "because we were like little sisters to them," explained Susie.

At some point, Guy developed a crush on Kathy. And when he was not quite eighteen, he made a gallant and rather old-fashioned gesture. He got up the courage to ask her father, Norm Nissen, for permission to ask his thirteen-year-old daughter out on a date. However, Norm turned him down. "My dad said, 'Noooo, you're too old,'" explained Kathy. It was another thirty years before Guy got up the nerve to ask her out again.

In 1965, when Guy was fourteen years old and Rich was just thirteen, a golfing buddy of Harry Snyder's convinced him to invest in a 50 percent stake in a new drag strip called the Irwindale Raceway. Built in 1964 with a quarter-mile track, the "Dale" (as it came to be known) was one of the numerous sanctioned drag strips sprouting up in Southern California in the wake of the founding of the National Hot Rod Association (NHRA). The investment soon proved to be a canny move that further solidified the association between cars and In-N-Out that began with the drive-through. It was also a natural extension of the time when In-N-Out was a stop on the illegal street-racing circuit.

Born in the salt flats and dry lakebeds of the Mojave Desert during the 1930s—where hot-rodders first topped speeds of 100 mph—drag racing had changed greatly in the postwar years. Where dragsters once had free rein to run the lakebeds at El Mirage and Lake Muroc, they increasingly found themselves edged out by the U.S. Air Force. Taking advantage of the temperate climate and the flat, vast, and remote lakebeds, the air force began testing new jets and rockets there. On October 14, 1947, Chuck Yeager broke the sound barrier while flying the experimental Bell X-1 in the sky over Lake Muroc at Mach One at an altitude of 45,000 feet.

A year later, Alex Xydias, a B-17 engineer during the war and no stranger to the lakebeds, opened the So-Cal Speed Shop on Olive Avenue in Burbank. It was at El Mirage that Xydias hit a record 130.55 miles per hour in his belly-tanker made from a Ford V-8 60-horsepowered engine and a simple open-wheel aluminum frame. Soon, the lakebed days were numbered. In 1949, Muroc was officially converted into Edwards Air Force Base and became the testing ground for X-15s and later the landing site of the space shuttle. The dragsters were forced to move their action from the salt flats to urban asphalt.

While some of the contests were held on abandoned airstrips, the majority of them were held on city streets, luring crowds and creating a public nuisance. Law enforcement took a decidedly dim view of the dragsters, and had long sought to clamp down on the illegal street scene, implementing ordinances against hot-rodding and closing down streets altogether.

For many young men, racing powerful, noisy cars at maximum speed was not just a bullet of freedom but was a revolt against the parochial values of their parents. It wasn't just a sentiment among hot-rodders; it could be found among the automakers in Detroit as well. In fact, Clare MacKichan, one of the design engineers behind the seminal 1955 Chevy, later explained that the car's design target was to represent "youth, speed, and lightness"

In an effort to "create order from chaos" (as he put it), in 1951, Wally Parks, a dragster himself and the editor of *Hot Rod* magazine, launched the NHRA. A former military tank test driver for General Motors who had served in the South Pacific during World War II, Parks saw the long-term advantages of organized drag racing on off-road sites. In creating the NHRA, Parks realized that he could promote and legitimize the sport of legal speed racing while combating its low-rent, outlaw image. The NHRA implemented safety rules and performance standards and began operating a number of drag strips and races under its own sanctioned program. For a time, the NHRA sponsored a traveling Safety Safari, a portable drag strip caravan that promoted the sport across the country. As a result, drag racing became a hugely popular spectator sport, at times even eclipsing baseball and football games, and soon attracted commercial sponsors.

The sport had so transcended its humble beginnings that it was the subject of an April 29, 1957, *Life* magazine cover story: "Hot Rod Fever." By the early 1960s, the NHRA had over 130 approved strips in forty states under its sanctioned umbrella. The Dale was one of them.

Located just west of Irwindale Avenue and north of 1st Street, near the boundary of Irwindale and Azusa, the strip was perfectly primed to become a hotbed of Southern California's growing hot-rod activity. The city of Irwindale, a flat expanse dotted with rock quarries, small industrial parks, and the occasional battalion of palm trees, was just twenty miles east of downtown Los Angeles and a short three miles from Baldwin Park. Flat, dry, dusty, and ferociously hot in the summer, Irwindale had more rock quarries than residents. In fact, the city was nicknamed *Jardin de Roca* (Spanish for "Garden of Rocks") owing to the fact that nearly every highway in California and a number of those west of the Mississippi were made from Irwindale's rock and gravel pits.

At the time of Harry's involvement with the drag strip, there were only a handful of In-N-Out Burger outlets fanning out from Baldwin Park. Each was hugely successful, replicating the popularity of the original, claiming long meandering lines of near-rush-hour proportions from opening to closing time. But In-N-Out Burger was still something of a local phenomenon, clustered strictly within the San Gabriel Valley. The Dale opened up In-N-Out to a much wider audience.

Possibly owing to his early days working various refreshment and game stalls at the Venice pleasure piers, Harry Snyder shrewdly decided not to be a mere investor but to supply the food at the racing track as well. There were two concession stands; one on the north side where the starting line, staging area, and spectator stands existed, and the other on the south side where the pit and finish line stood. The stands sold In-N-Out burgers—but they weren't officially part of the growing chain. They were simple wooden shacks at the Dale with the words "Snack Stand" painted in black across them.

The racers and fans who frequented the Dale fondly recalled that the burgers and fries were almost as much of a draw as the races themselves. "I'm not sure if it was the atmosphere or what, but they tasted better there than anywhere," recalled Valerie Althouse, who raced at the Dale between 1969 and 1971. "I think people loved to come to Irwindale for the food. One of the reasons that the track had such a great reputation was basically [that] the snack stands were In-N-Out even though they didn't say so."

On any given race day, the Dale's grounds were swarming with cars and people. Young kids used to sneak into the track through the gravel pit that stood outside the staging lanes. In those early days—in the mid-1960s—the track operated races that ran all weekends from noon until ten at night as well as on Wednesday evenings. It was still the early days of drag racing, before the big money races, when the dragsters paid to get into money brackets—with winners taking home a whopping one hundred dollars and, of course, gleaming trophies. The Dale was so frequently packed and noisy that at one point residents in neighboring Azusa complained, asking the raceway to research noise control.

The rowdy races helped In-N-Out Burger earn a reputation outside of Southern California. Racers who had been to Irwindale began spreading the word about its burgers. As part of the NHRA, people from all over the country began descending upon the Dale and its legendary snack stands. Soon enough, going to an In-N-Out Burger became part of the racing circuit experience.

One of the first female dragsters, Eileen Daniels (who began racing in 1955), can still remember her first In-N-Out experience. It was 1957, and Daniels (along with her husband, Bob) was running the raceway in Indianapolis for the NHRA. On a trip to Southern California, "some guys from the NHRA told me I just had to try an In-N-Out Burger," she recalled. "I got the cheeseburger and the french fries. Just outstanding, great flavor." That was it. Going to In-N-Out Burger became a ritual. Actually, more than a ritual—it grew into a fifty-year habit. Daniels explained, "I make it my very first meal when I come to California."

During his years at the Dale, Harry Snyder opened a handful of new In-N-Out Burgers. Between 1966 and 1972, he unveiled new stores in Azusa, La Puente, Pomona, and even ventured outside of the chain's home turf, launching new shops in North Hollywood and Panorama City in the San Fernando Valley (for comparison, by 1968* there were one thousand McDonald's across the country).

Harry often conducted business meetings for the Dale and In-N-Out from his San Dimas living room. Esther left Harry to be the public face of their ventures. She took care of the bills and handled the paperwork of both businesses—but while she preferred to remain behind the scenes, nothing escaped her knowledge. "She was a great lady, just incredible," remembered Steve Gibbs, whose childhood relationship to In-N-Out Burger came full circle when he was tapped by Harry to manage the track in 1966. A veteran of the San Gabriel drag racing circuit, Gibbs grew up in Baldwin Park and courted his wife during high school in large part at In-N-Out Burger.

Gibbs was just twenty-six when Harry offered him the top position at the track. He had been working part-time initially, just on the weekends, and after Harry noticed Gibbs and his hard work, Snyder increasingly gave him more to do before making him manager. "Harry had strong opinions on a lot of things but he wasn't hard to work for—you knew where he was coming from. He was clear." Gibbs, who would later become the vice president of competition for the NHRA, also remembered Harry's fiery temper and his ability to cool down and let bygones be bygones, "he could blow up, then say his piece over it and he'd be laughing later on," he said. "If he was upset, he'd tell you why. He was probably right, but if you had a good answer, he'd accept it."

While they were still just in junior high school, Guy and Rich Snyder went to work at the track, too. They did odd jobs like running

---

* 1968 was the year that McDonald's introduced the Big Mac nationally.

the elapsed time slips to the racers or cleaning up trash around the track. "They weren't spoiled little kids running around like you'd expect, being the owner's sons," Gibbs recalled. "They were good kids. I never had to get too hard on them. They did their work. That's the way I remember them. The Snyders had a strong work ethic and I think they wanted that for their boys too."

Intense, dogmatic, and hardworking—this is how Harry Snyder is invariably described. At the same time, he displayed an unwavering sense of decency. "When we had a good race he'd give us bonuses," recalled Gibbs. "He was very conscientious about being fair to customers. You saw that at In-N-Out, and at the track, too. I remember one time we put on this big event that was above and beyond the weekly races, and it was costly. We had to raise ticket prices. Harry agonized over this. At the time, tickets cost $2 for a Saturday night, and he had to go up to $2.50 or $3. It sounds like peanuts now, but then it was a big decision. Harry settled on $2.50. He didn't want to be too hard on the weekly regular customers, and he felt that $3 was too big a jump. Besides, we were already making good money." Gibbs added, "I really respected him. He was a good, solid businessman, and he treated people well and tried to do the right thing."

Perhaps as a result of Harry's brief flirtation with communism and his disgust with economic inequity, he always exhibited a soft spot for those in need. He had a real eye for potential and took a special interest in developing those individuals who worked for him. While Steve Gibbs managed the raceway, Harry paid for him to take a Dale Carnegie course. (Carnegie, of course, was the early American guru of self-improvement, salesmanship, and corporate training, and the author of perennial best seller *How to Win Friends and Influence People*.) "He was always trying to make things better," recalled Gibbs. "I was a blue-collar guy. You know he didn't have to that."

It was during this time that Harry Snyder met Paul Althouse at the burly dragster's San Dimas shop—Harry was impressed enough to make him an offer. Althouse (husband of Valerie Althouse, another Dale dragster) had grown up in the trailer parks of Baldwin Park and on In-N-Out burgers. By the time he met Harry, the veteran Southern

California racer and master of the four-speed had also made a name for himself fixing cars and building innovative street hot rods that were fast and affordable. The '65 Chevelle he built for Geno Redd was one of the first to have a 375-horsepower 396-engine; the Chevelle went on to set a record time of 11.54 seconds, and win a major race at the Lions Dragstrip in 1965.

Every three or four months, Harry made an appearance at Althouse's shop to have some work done. Their initial interactions were simple enough. "At first, I didn't even know who he was," Althouse claimed. "I knew the boys who worked at the track." Then one day, seemingly out of the blue, a teenaged Guy Snyder came to him and said, "My dad wants to talk to you." The two men knew each other slightly at best, but Harry asked Althouse if he would like to open a Tuneup Masters on his property. "He offered to finance it 100 percent," said Althouse. It was an extraordinary opportunity, and Althouse was as surprised as he would have been if he had been told he'd won the lottery. Still, he chose to decline the offer. "I told him it sounds like a good deal to a point, but to be perfectly honest, I don't have the education to do it and do it properly. He very much appreciated that." Nevertheless, Althouse remained within the Snyder family's orbit for several decades after that first meeting in the family's living room. Said Althouse, "Harry was the kind of man that if he believed in you, there was nothing he wouldn't have done."

Harry used to say, perhaps only half-jokingly, that one of the reasons he had invested in the Dale was to keep his sons off the streets and out of trouble. For the Snyder boys, their jobs at the Dale became more than just work and a way to earn pocket money. Those days at the track became a way of life. Like the In-N-Out Burger stand, the raceway became the Snyders' playground. Working the snack stands and being around the cars and racing became more than just a hobby for Guy and Rich; it became a lifelong pursuit. The brothers raced as soon as they were old enough, and they began collecting and fixing up cars: muscle cars, funny cars, old Willys, and roadsters. It was during those days at the track that the Snyder brothers' primary interests took root and their futures took shape. At the raceway, Guy's singular love affair with drag racing and Rich's enduring love for running In-N-Out were born. As Steve Gibbs explained, "The boys got the bug, particularly Guy. He was a real racing nut, but Rich, he took the company."

Around 1972, Harry sold his interest in the raceway. Five years later, the Dale was shuttered for good and the tracks were plowed under to make way for a giant Miller Brewery. Friends said they thought the track had never been more than a fun and profitable sideline for Harry and that it was his burger chain that was always his first priority. Although Harry Snyder ended his formal business relationship with the Irwindale Raceway, that did not necessarily signal the end of In-N-Out Burger's involvement with hot rods and drag racing. If anything, the Irwindale days were just the beginning of a lifelong association between the Snyders and dragsters. But for now, Harry's full attention turned to In-N-Out Burger. There were about ten In-N-Out Burger locations, and the independent chain was not only growing but prospering. At the same time, the little chain was facing some serious issues concerning the future direction of the company. While In-N-Out operated exactly as it had on the first day it opened for business in 1948, the same could not be said for the rest of the fast-food industry.

CHAPTER 7

Harry shrugged his shoulders at the rapid changes taking place all around him and quietly dug his heels into the ground. Fast food was spreading at a fevered pitch, cementing its place in American culture. Before long it made up a major piece of the nation's economy as well. As it happened, the dawn of what came to be thought of as the all-American meal collided with a new era in American history: the Nuclear Age. On October 4, 1957, the Soviets launched *Sputnik*—the first satellite to orbit the earth—ushering in the space race. Back on earth, the public had become fascinated with technology. A belief in science ruled the day, and it was characterized by a veneration of shiny new machinery and the magic that it could perform. Bell Labs began to create artificial intelligence, and scientists developed a crude network of computers that linked government agencies and universities that later came to be known as the Internet. There were other marvels of modern science with a more immediate impact; on April 12, 1955, Dr. Jonas Salk announced to the world that he had discovered a vaccination for polio that was "safe, effective, and potent."

The idea that technology could enhance the everyday lives of American citizens was widely circulated. It was, after all, during this time that the Eisenhower administration came up with its "Atoms for Peace" campaign, a propaganda effort aimed at bolstering the image of nuclear energy in the public's mind. It was none other than Walt

Disney who was tapped to popularize the drive; Disney's amusement park as well as his publishing, film, and television empire had made him, according to *Time* magazine, "one of the most influential men alive." And in 1957, Disney (in conjunction with the U.S. Navy and General Dynamics, manufacturer of the nuclear submarine, USS *Nautilus*) produced a live-action/animated feature called *Our Friend the Atom*.*

The wealthier postwar American society increasingly turned to science and technology in order to solve its problems and make daily life easier. Furthermore, new gadgets and gizmos were no longer the purview of the rich. Teenagers bought cheap transistor radios and record players; their mothers enjoyed a host of new appliances in their kitchens. Automatic electric washers and dryers, steam irons, electric can openers, and electric coffee pots all offered more than just time-saving convenience—they promised perfection, too. It was only a matter of time then before modern technology filtered down to how Americans ate, the way their food was prepared, and even what it was that they were consuming.

Emblematic of this radical new culinary trend was the introduction in 1953 of the frozen "TV Dinner" invented by a C. A. Swanson & Sons of Omaha food technologist named Betty Cronin. The prepackaged meal of turkey and potatoes (sold for ninety-eight cents) could go from the oven to a TV tray—and when finished, just as swiftly and neatly to the garbage can. Swanson's "Quick Frozen Turkey Dinner" was an immediate hit. Initially expecting to sell five thousand units, the company actually sold 10 million meals during the TV Dinner's first year on the market.

Like Cronin, food technologists across the country were spending time in their kitchen laboratories and hatching all sorts of new ways to process, mimic, reduce preparation, speed up cooking, and re-

---

* In 1959, General Dynamics constructed Disneyland's fleet of "atomic" submarines for its $2.5 million Submarine Voyage attraction. It was one of the park's first e-ticket rides.

constitute everything fit for human consumption. During the 1950s, when Americans snapped up over $1 billion worth of frozen foods, a number of food-processing firsts emerged. The Seeman Brothers of New York introduced the world's first instant iced tea, the White Rose Redi-Tea, in 1953. A year later, as the prices for cocoa spiked, Robert Welch, a Boston candy maker, gathered a group of chemists and came up with a less expensive but satisfactory replacement. In 1959, General Foods Corporation began marketing a powdered, non-carbonated, orange-flavored beverage called Tang.

It was the era of easy, instant food. And nowhere did this budding emphasis on the frozen, dehydrated, prepackaged, and automated find a more welcoming home than the growing fast-food industry.

Leading the charge was McDonald's. Under Ray Kroc, Dick and Mac McDonald's basic principle of a high-volume, low-cost operation had became nothing less than a string of large and efficient food manufacturing systems. The infrared heat lamps deployed by the McDonald brothers in the early days to keep pre-made burgers and fries hot and ready and their multiple meat patty molding devices gave way to a whole slew of innovations. Kroc hired engineers and technicians dedicated solely to upgrading and mechanizing the McDonald's process. There was the "clamshell grill" that sliced cooking time in half by cooking both sides of the meat patty simultaneously, the deceptively simple V-shaped aluminum scoop with a funnel at the end that could pack a bag of fries neatly in one fell scoop. Each order required little more than a quick assembly from a ready-made food kit.

McDonald's was not alone. Most of the leading chains looked for new and cheaper ways to keep up with the staggering output of food; as a result, entirely new industries devoted to the manufacture and processing of food and artificial flavors mirrored in no small measure the growth of the fast-food industry. Computerized timers were deployed to set off alarms when it was time to flip burgers, conveyor belt flame-broilers cooked frozen patties into hamburgers in under sixty seconds. Cybernetic deep fryers automatically cooked fries to perfection, and ketchup and mustard were applied to buns in uniform amounts by electronic pumps.

As early as 1931, White Castle, the Wichita-based chain, replaced fresh beef with frozen square beef patties. Some ten years later, the chain came up with a time-saving system of punching five holes in each hamburger to facilitate quick and thorough cooking so that the patties didn't need to be turned. But it wasn't until Ray Kroc made the shift from fresh to frozen that nearly the entire industry followed.

In 1968 there were about one thousand McDonald's restaurants across the country pulling in $355.4 million in sales. To supply them all, the company was using 175 different meat suppliers. After twelve years of insisting on fresh beef, reluctantly, and after much internal opposition, Kroc made the switch. He agreed after a small, family-owned Pennsylvania meat de-boning firm called Equity Meat Company had proved that it could standardize its system of cryogenically freezing beef patties that didn't leave the meat desiccated and taste-less. Equity made it possible to preserve frozen patties for extended periods of time. With few exceptions, this proved to be a point of no return for the rapidly expanding fast-food industry.

Soon, Equity was converted from a strictly beef-boning operation into a manufacturing facility. With McDonald's as its lead customer, Equity (later renamed Keystone Foods) became the world's largest hamburger supplier. By 1970, the company reported manufacturing 1.5 million pounds of hamburger annually in the United States. Two years later, that number rose dramatically to 50 million pounds.

Coincidentally, the same year that Betty Cronin invented the TV dinner, John Richard "J. R." Simplot, the country's largest supplier of fresh potatoes, devised a canning and dehydration quick-freeze plant that could freeze french-fried potatoes on a mass scale. More important, Simplot's frozen product was nearly indistinguishable in taste, color, crispness, and uniformity from the fries made from fresh potatoes. In 1965, Simplot, a tough-as-nails Idaho entrepreneur and farmer, met with Kroc and convinced him to buy his quick-freeze potatoes. They were both self-made men, and each had made a for-tune on their own terms. Simplot seemed to understand that there was really only one point on which he could persuade Kroc. As he later recalled, "I told him frozen fries would allow him to better control the quality and consistency of McDonald's potato supply."

Before long, Simplot became the frozen french fry supplier for such outfits as Burger King, Wendy's, and Kentucky Fried Chicken. (Soon, food conglomerates such as Lamb Weston and Carnation would provide other chains and restaurants with their flash-frozen french fries.) From frozen, processed potatoes, J. R. Simplot went on to build an estimated $3.6 billion empire on devising "labor-saving, value added foods," for commercial use.* The shift from fresh to frozen was only one part of the changing fast-food equation. In 1946, the U.S. Department of Agriculture required that the fat content in ground beef classified as hamburger not exceed 30 percent. As a result, meat suppliers filled in the remainder with a number of cheaper non-beef additives and meat extenders such as soy protein that could absorb moisture and reduce cooking time. Low commercial value parts of cattle like tripe and cheek were ground into the beef while nitrates were used to enhance the meat's pink color. As Jim Williams (president of Golden State Foods Corporation, a large California hamburger supplier) told author John Love, "You would negotiate a price with the drive-in and then find a way to make it work economically. All-beef hamburger was a myth. There were very few additives that weren't used."

During these years, an entirely new field of artificial and chemical food additives developed, and the exploding fast-food industry was beginning to rely heavily on them. In *Fast Food Nation*, author Eric Schlosser's polemic on the industry, he lists page after withering page of the numerous chemicals, additives, and coloring agents lacing the leading chains' salads, soft drinks, condiments, and sandwich buns. In one particularly appetite-crushing passage, Schlosser listed forty-seven chemical ingredients used to make the strawberry flavoring found in a Burger King strawberry milk shake. The relentless competition and new preparation techniques had transformed the entire industry that had once uniformly prided itself on the personal touch.

---

* Simplot died on May 24, 2008. A year earlier, at ninety-eight years old, Simplot was listed as the 214th richest American (and oldest living billionaire) on the *Forbes 400*.

Harry Snyder was having none of it. Over in Baldwin Park, things looked much the same as they had when artist Jack Schmidt painted the original In-N-Out Burger stand in 1948. Harry was determined to do things exactly as they had always been done. He had developed and clung to his own values, focusing on quality products and quality people. Harry had clarity about his decision to stick with what by the late 1950s was already beginning to appear old-fashioned.

Around 1961, Harry's nod to advanced technology was to switch from gas to electric grills—but the In-N-Out hamburgers still required the human touch. It was the stores' managers who continued to flip the chain's burgers. There were no infrared heat lamps or microwave ovens, no prepackaged or frozen food at Harry's drive-through. All burgers and fries were made to order. To help cope with the increased output of burgers, Harry added a second grill to the original kitchen in Baldwin Park.

To Harry's mind, there was just no substitute for the real thing. All hamburgers were made from fresh, 100 percent additive-, filler-, and preservative-free beef. By 1963, as many of the leading chains were beginning to use frozen beef patties,* Harry Snyder hired In-N-Out's first butcher. He wanted to exert more control over his products, not just find a way to wring more money out of cheaper products. At one point, he purchased a ranch and hoped to raise his own beef, but found it just wasn't cost effective. Instead, Harry built a facility near store Number One in Baldwin Park where his specially selected chucks were delivered. There In-N-Out's butcher boned, hand-cut the chuck's front ribs and shoulder (Harry insisted that no other part of the steer was ever used), ground it up into beef, and molded it into hamburger patties before delivering them fresh to each store. "This way," the company later proclaimed, enabled it to "completely control the patty-making process." While it appeared

---

* In an exception for the large chains, Wendy's Old-Fashioned Hamburgers would use fresh, not frozen, beef patties.

that In-N-Out Burger was either fanatical, staying behind the times, or both, Harry's decision proved to be a critical moment in deciding just how the company would define itself.

In-N-Out took the same approach with its french fries—the Snyders still made theirs by hand. Burlap sacks of whole, fresh Kennebec potatoes specially grown for In-N-Out arrived in Baldwin Park and were distributed to each store where they were washed, peeled, cut, and cooked in cholesterol-free, 100 percent vegetable oil. Frequently, the potatoes were picked in the morning and delivered to In-N-Out the same evening. Harry made sure to scrutinize the freshly delivered sacks. They were inspected for starch content and a test batch of fries was made up right away. If the potatoes weren't up to muster, the whole truckload was rejected.

In an industry that was substituting chemically processed, prepackaged, and frozen food for the real thing, In-N-Out continued to use traditional sponge-dough buns, fresh-baked daily, that took several hours to rise—its competitors took to purchasing buns injected with chemicals that considerably reduced the rising time of the dough. As others began using cheaper and less labor-intensive concentrated ice milk mixtures (with preservatives that could be frozen later and reconstituted into milkshakes), In-N-Out refused to use anything but 100 percent ice cream in their milkshakes.

By the time that Harry had opened about half a dozen In-N-Out Burgers, television had eclipsed radio's longtime dominance. TV ownership had grown exponentially, from 3 million households in 1949 to two-thirds of all American homes in 1955. If the American public had come to view the fast-food hamburger as the food of a new America, this belief was soon popularized by television commercials that could blanket the message over the widest territory possible.

Until the early 1950s, most American restaurants relied largely on word of mouth. Few if any advertised beyond placing an ad in the local Yellow Pages or announcing their establishments on billboards placed alongside highways. But the explosion of fast-food chains was accompanied by new and aggressive marketing campaigns, and the advertising of mass consumption was amplified by the strikingly effective mass medium of television.

McDonald's launched its first national commercial in 1967. The $3 million campaign had the catchy jingle "McDonald's is your kind of place." Before long, pervasive campaigns dominated the airwaves. They capitalized on images of American family life; later, they focused on idealized experiences and created alternative universes with fuzzy mascots that were heavily geared toward children. Ronald McDonald, a clown in a bright yellow jumper, big rubber feet, and a bulbous red nose, made his debut in 1963. He was soon joined by a host of characters like Mayor McCheese and the Hamburgler who lived in the fictional McDonaldland, a magical world filled with milkshake volcanoes and apple pie trees.* Jack in the Box, founded in 1951, initially used a toy clown in a box as its mascot and later a cute kid named Rodney Allen Rippey. In 1955, Burger King began featuring its own Magic Burger Kingdom with a king who sat on a burger throne. The advertising campaigns had storylines like Saturday morning children's shows, they had jingles and slogans that millions of consumers knew by heart. They focused on fun and fantasy and, for a time, seemingly everything but the actual product.

The story at In-N-Out remained static; the message was hamburgers and fries, and it was broadcast by its customers. The chain's simplicity and old-fashioned values certainly endeared In-N-Out to its longtime regulars, establishing an uncommon customer loyalty—one that its competitors spent hundreds of millions of dollars trying to create.

In fact, In-N-Out's bond with the public took on a life of its own, and the chain's desire to please its guests remained its primary goal.

---

* In 1973, television producers Sid and Marty Krofft successfully sued McDonald's and its advertising agency for copyright infringement, claiming the McDonaldland characters were a rip-off of the Kroffts' own popular children's show *H.R. Pufnstuf*. The judgment was decided in 1977.

If a customer wanted her meat bloody and raw, that's what she got—bloody and raw. If he wanted extra pickles, or his meat well-done, or his fries extra salty, no problem. Soon, variations on the standard bill of fare turned into what regulars began calling the "secret menu." Customers ordered these variations with such frequency that a number of them were given names. Favorites included: the 4x4 (four beef patties and four slices of cheese); the Flying Dutchman (two beef patties and two slices of melted cheese—no bun, produce, or condiments); Animal Style (mustard-cooked beef patty served on a bun with pickles, lettuce, extra spread, and grilled onions); Protein Style (burger or cheeseburger wrapped in lettuce without a bun); and Grilled Cheese (no beef patty).

Before long, the secret menu became something like a secret handshake that was traded among regulars and their friends—but In-N-Out was in on it as well. It became common knowledge among the chain's associates, who understood immediately when a customer asked for, say, a cheeseburger Animal Style.

The secret menu added greatly to the chain's growing mystique. Since its origins are untraceable, the secret menu developed its own peculiar folklore. Before long, various narratives became associated with the different styles that developed. In one version, Animal Style was created when surfers and hippies started ordering their burgers with mustard cooked into the beef—and the clean-cut associates started calling it Animal Style as a joke. Others decided that customers had given it the name Animal Style because it was messy to eat. In one telling, Protein Style emerged because Harry Snyder (while on a diet) ate the burger with lettuce instead of a bun. And the Flying Dutchman was reportedly Guy Snyder's favorite. Its name, of course, was also a nod to the family's Dutch roots.

Just as In-N-Out Burger was the antithesis of McDonald's, Harry Snyder was in many ways the opposite of Ray Kroc. The flamboyant Kroc had encouraged publicity at every turn. Very early in the game, he hired publicity teams who succeeded in getting stories on both

Kroc and McDonald's published in newspapers across the country. And long before Sir Richard Branson injected himself into stunts dreamed up to advance Virgin's visibility (like traveling across the Atlantic Ocean in a hot air balloon), Kroc had been known to pull a few tricks from his own leisure suit pocket. In the mid-1960s, McDonald's traced the origins of the hamburger to a Russian sailor in the German port city of Hamburg and journeyed there to present the mayor of Hamburg with a McDonald's hamburger.

McDonald's was not the first to deploy gimmickry, however. In 1930, White Castle founders Billy Ingram and Walter Anderson came up with a scheme to counter the public's squeamishness about eating ground beef that endured long after the publication of Upton Sinclair's 1906 meatpacking industry exposé, *The Jungle*. Ingram paid a group of young men to dress up as doctors while they ate White Castle hamburgers to promote the idea that hamburgers were healthy.

Harry had little interest in such attention-grabbing tricks or publicity stunts. Esther's modesty and her husband's salt-of-the-earth soul recoiled from the kind of promotional splashes going all around them. Their own promotions were small-scale, such as producing wooden nickels redeemable for soft drinks. Their philosophy was simple; the product—if it's a good one—should sell itself, and everything else is smoke and mirrors.

Harry and Esther believed in serving the communities in which In-N-Out operated. The Snyders made any number of charitable donations, and their efforts at promotion were often connected to grassroots community and philanthropic endeavors. The Snyders served their burgers at schools and little league games for free or at a significant discount, and they offered their own expertise and help whenever Baldwin Park's Adult School asked. Esther was particularly interested in helping children. The Snyders offered In-N-Out door prizes at community-sponsored events. The result was that In-N-Out generated a kind of homespun feeling; there was a consistency and authenticity about the chain.

While the couple had long since moved to the more prosperous suburb of San Dimas, they remained very much a part of Baldwin

Park. Allen Teagle, who returned to Baldwin Park with his wife, Bonnie, following his stint in the navy, put it this way: "Everyone knew who Harry and Esther were. Whenever we had a Lion's Club event, Harry always donated the food. That's how we raised money."

Harry was active in the Baldwin Park Rotary Club, the service organization established in 1905 by Chicago lawyer Paul P. Harris that brought together business and professional leaders to provide humanitarian service, encourage high ethical standards, and build global good will. The organization's motto, "Service Above Self and They Profit Most Who Serve Best," was an apt description of Harry and Esther's own beliefs.

At Rotary Club functions, the Snyders often provided In-N-Out burgers. Joe McCaron, whose father, Joe Sr., was Baldwin Park mayor from 1968 to 1970, remembered one Rotary-sponsored occasion during his father's term, when families gathered for a barbecue and the kids were given fancy saddles and rode horses along the San Gabriel River. "The Snyders brought coolers of In-N-Out meat and fixings and we had a barbecue," recalled McCaron, who remained close to the family and worked with In-N-Out years later servicing their cookout trailers. "Nobody ever had barbecue In-N-Out burgers, they didn't do that, and I'm not sure they've done it since."

When asked to account for the chain's success, Esther Snyder once said that it had been "accomplished only with the dedicated enthusiasm and wholehearted cooperation of the In-N-Out Burgers employees and our pleased customers."

CHAPTER 8

In these years, it was easy to see what made In-N-Out Burger a beloved, cult phenomenon; it was its unyielding simplicity and dedication to quality and fair practices, the principles that Harry and Esther had put into play from the start. The Snyders built In-N-Out by continuously producing quality hamburgers and fries, reinvesting in their employees, and keeping the chain's growing legion of customers happy: nothing more, nothing less. In-N-Out's enduring success stemmed from Harry's capacity to understand what he did best and focus exclusively on it. In part, it was what a couple of Bain & Company consultants would years later term "the innovation fulcrum"—that is to say, identifying the point at which introducing new products or services actually destroys more value than it creates.

Harry's business acumen had nothing to do with fancy management consultants or popular buzzwords and everything to do with his own core belief: "Keep it real simple," he said repeatedly. "Do one thing and do it the best you can." And he stuck to his word.

As the years wore on, in the minds of In-N-Out's longtime regulars, there was something reassuring about the little red and white drive-through restaurant that continued to make their hamburgers from fresh ground beef and served its fresh, hand-cut fries in a paper boat. As a result, a deep attachment developed between customers and the chain that they could always count on. As In-N-Out grew, that relationship grew along with it. "In-N-Out always tasted the same," explained Elaine Setterland, a high school classmate of Guy and Rich Snyder's. But it was more than that, she said. "In-N-Out was local, it was ours." Even as the chain grew and moved beyond Baldwin Park, it retained that sense of local intimacy. "In-N-Out was a way of life," proclaimed Steve Gibbs. "We just kind of took it as part of us. We grew up with it. It's funny, when our friends get together for high school reunions, In-N-Out invariably comes up, even after fifty years."

Harry resisted opportunities to franchise or to cash out. Despite the enormous pressures and the options surrounding him, Harry rejected the conventional wisdom of the industry. Inside the company, franchising was a dirty word. In building In-N-Out Burger, Harry followed no compass but his own. There was no hierarchical management structure, no bureaucracy, and there were no shareholders to answer to. There were just Harry, making decisions from his gut, and Esther, standing by his side, looking after the books. At the start of the 1970s, a time of extraordinary growth and change within the fast-food industry, Harry could spread out a map of his small burger kingdom and smile. In 1973, after twenty-five years, there were only about a dozen In-N-Out Burger outlets. He owned each store and the land underneath them. The company's debt was negligible—and that was exactly how Harry liked it.

What had started as a kaleidoscopic assembly of little family-run roadside stands in the 1930s and 1940s had, by the early 1970s, been transformed into a booming nationwide industry increasingly run by corporate titans. In 1973, more than 245 franchise companies had launched more than 32,000 fast-food establishments that sold $9.68 billion worth of hamburgers, hot dogs, pizzas, chicken, tacos, seafood, donuts, pancakes, and sandwiches. The all-American meal had become the American way. Fast food was part of the great free enterprise system, and its corporate masters were spreading their message one french fry at a time. As the folksy host of CBS's *On the Road* Charles Kuralt once remarked, "You can find your way across this country using burger joints the way a navigator uses stars."

During this period, the fast-food industry was growing at an average rate of about 15.5 percent annually. It was hardly a coincidence then that the big three—McDonald's, Burger King, and Kentucky Fried Chicken—were each transforming into veritable empires. Between 1965 and 1971, Kentucky Fried Chicken went from one thousand to four thousand units. Burger King grew from one hundred to eight hundred stores during the same period. Not a decade earlier, Ray Kroc had boasted that McDonald's would open one hundred new restaurants a year; now the chain was opening one new store each day. Between 1965 and 1973, the number of McDonald's outlets in the United States alone rose from 738 to 2,500. A host of imitators proliferated in their wake—many succeeded, and many more failed.

With its promise of a quick, self-contained meal to go served by well-scrubbed teenagers in paper hats, fast food was not just the all-American meal; it had become emblematic of American life. The now telltale squat structure with a mansard roof and a bright neon sign had become as common a sight as churches, schools, or supermarkets. Broadening their scope, the chains moved from downtown city centers and highways to suburban areas situated near both commercial districts and residential neighborhoods. That way fast-food restaurants could capture the working lunchtime crowds during the day as well as the traffic made up of families going out for an affordable dinner in the evenings. For a time, the opening of a new fast-food restaurant, particularly in small towns, was a cause for celebration.

In 1971, Harry opened up a new In-N-Out Burger drive-through on Lankershim Boulevard in North Hollywood. It was his first move beyond the San Gabriel Valley, and he chose wisely, pushing into the growing San Fernando Valley. Following World War II, the 345-square-mile Valley, located in the northwestern section of Los Angeles, had mushroomed into a major bedroom community. Between 1945 and 1960, the Valley's population nearly quintupled from 176,000 to 840,000, growing into a large quilt of middle-class homeowners with swimming pools and two-car garages that earned it the nickname (coined by a *Los Angeles Times* reporter) "America's Suburb."

Despite countervailing trends in the industry, Harry still positioned In-N-Out for the car-reliant customer, and the new North Hollywood drive-through was close to the Hollywood Freeway, an important shortcut through the Cahuenga Pass that transported thousands between the Los Angeles Basin and the San Fernando Valley every day. Eminently visible and highly trafficked, the new In-N-Out store was also primed to take advantage of its proximity to Hollywood proper and its movie studios, continuing the early association between the burger chain and celebrities. According to Linda Hope, her father, Bob, was particularly thrilled when In-N-Out Burger opened in North Hollywood; the new store was close to the comedian's Toluca Lake estate, and he used to regularly surprise the associates working there. "He'd go through the drive-through and talk to the kids about golf," his daughter explained. "He just loved talking to the kids who worked there. The kids would say 'Do you know who was just here? Bob Hope!'"

The new store quickly became a magnet for Valley residents, who had heard about the chain from expatriate Baldwin Park residents or experienced In-N-Out on trips through the San Gabriel Valley. By the mid-1970s, there was already a first generation of In-N-Out fans who had grown up eating the chain's burgers and fries. As they achieved their majority and moved outside of Baldwin Park, they took their devotion to In-N-Out with them. They became the cornerstone of what had become a strong word-of-mouth campaign that helped to fuel desire for In-N-Out, establishing a grassroots marketing drive. The growing freeway system had abruptly merged the isolated pockets of communities that jutted out from downtown Los Angeles. It made it possible for a small chain like In-N-Out to broaden its appeal simply by watching its customers travel up and down California's network of freeways, spreading the good word.

Perhaps nothing exemplified the industry's pervasiveness in American culture better than its most dominant player, McDonald's. For years, the company had proudly proclaimed the millions of burgers it sold. And in August of 1973, the company raised a neon banner on its now ubiquitous golden arches that read: "Over 12 Billion Sold." With more than $1 billion in sales annually, McDonald's surpassed the U.S. Army as the country's largest distributor of meals. When the chain launched its $50 million "You Deserve a Break Today" campaign in 1971, the catchy theme fast became one of the most identifiable jingles of all time.

Fast food, like IBM or General Motors, was big business—and Wall Street eagerly invested. In 1965, McDonald's went public with the company's first stock offering at $22.50 per share. Over the next thirty-five years, the stock split twelve times. In 1969, the year of Taco Bell's IPO, Kentucky Fried Chicken also went public. KFC (as it later became known) was listed on the New York Stock Exchange. When Carl Karcher took his company public in 1981, he raised $13.8 million.* It wasn't long before corporate America came calling. A handful of acquisitive food conglomerates and packaged food corporations, intent on taking advantage of the growing fast-food business, went on a buying spree. In 1967, the Pillsbury Company—originally one of the largest producers of grain in the country—acquired Burger King for $19 million. Ralston-Purina, best known as a maker of breakfast cereals and pet foods, bought Jack in the Box in 1968 in a stock swap. That same year, the food conglomerate General Foods Corporation purchased Burger Chef for $16.5 million. A successful Indianapolis-based chain founded in 1957, Burger Chef was known for its low-priced, flame-broiled burgers and value combos.

By 1970, Kentucky Fried Chicken had made 130 of its franchisees millionaires. The power that the fast-food industry wielded was enormous. When, in 1967, McDonald's announced it was going to increase the price of its hamburgers from fifteen to eighteen cents, it forced numerous hamburger shops to slash their own prices or risk losing business.

It wasn't just Wall Street that stood up and took notice; economists, business schools, and the press followed the fast-food industry and its major players closely. In September 1973, *Time* magazine put McDonald's on its cover with the headline: "The Burger That Conquered the Country." In the pantheon of celebrated American businessmen, Ray

---

* Burger King waited until May 17, 2006, before going public. By then, it had a market capitalization of some $2.1 billion, and the company raised $489 million.

Kroc was hailed as one of the country's greats. Like Kroc, a number of fast-food titans had not only become icons of American entrepreneurialism but had also captured the public imagination as larger-than-life characters. They played an increasingly public role in their chains. The image of Colonel Sanders in his southern gentleman's white suit, goatee, and black bolo tie became internationally recognized as the symbol of Kentucky Fried Chicken.

Harry Snyder had little interest in media adulation. From his perspective, In-N-Out was simply a different creature than its competitors. With its fresh ingredients and custom orders (that took an average of twelve minutes from grill to delivery), In-N-Out barely fit the description of what was considered a fast-food restaurant. Quietly, as the 1970s unfolded, Harry continued to set In-N-Out on a course that would put it in direct opposition to the industry that was mushrooming all around him.

Around 1973, some ten years after Harry installed the chain's first on-site butcher, he opened the chain's warehouse and purchased its first semitrailer. The new commissary allowed the Snyders to ensure strict quality control even as the chain continued to grow. With the meat patties prepared there, the buns baked at the facility, and the lettuce, tomatoes, onions, and potatoes delivered and inspected there, In-N-Out could distribute (by refrigerated semi) its fresh ingredients to each store daily even as the chain moved to extend its geographic reach. It proved to be a winning strategy that worked in tandem with Harry's slow-growth model.

Also around 1973, Harry purchased In-N-Out's first mobile cookout trailer. Gleaming and portable, the trailers were small In-N-Out Burgers on wheels. They contained refrigerators and grills and could be rented for parties and events. When a store lost power or needed a supplemental grill, the trailers could be dispatched to just about anywhere. The net effect of the trailers was that they also helped to broaden the geographically limited chain's footprint while providing a moving billboard as they traveled the Southland. In the years to come, hiring an In-N-Out trailer for a private event would come to be seen as sign of cool all over Southern California.

On the surface, Harry's moves were all deceptively simple. They weren't meant to radically alter In-N-Out's operations or hasten its ability to grow fast and far. Rather, Harry put into play a series of initiatives that could efficiently maintain the chain's traditional standards of "Quality, Cleanliness, and Service" in a rocket-fueled fast-food world.

The rampant expansion that marked the beginning of the previous decade had, by the early 1970s, begun to reach its saturation point. The big companies figured out that they could apply capital (theirs and that of their franchises), corporate management, and distribution systems and thereby blanket markets. That in turn would allow them to advertise nationally and efficiently. Soon, chains were locked in competition to secure prime locations and penetrate market share on the local, regional, and national levels. Two years after General Foods purchased Burger Chef, the chain swelled rapidly to twenty-four hundred locations in forty-three states and overseas before its operations began to sputter. In 1972, General Foods took a $75 million loss on Burger Chef, and a decade later sold the chain to Imasco for $44 million, which converted many of the restaurants into the Hardee's chain.

Many of the assets that made fast food attractive to consumers and investors alike began to implode. The recession tied to the 1973 oil crisis battered the industry. Stock prices plummeted and many concepts failed. McDonald's actually picked up the pace of its rollout during the economic downturn to ensure its primacy following the shakeout. Since In-N-Out never followed the strategies or trends of its competitors, it was barely affected by the cyclical turn of events that first catapulted fast food to success and then savaged the industry. The Snyders' choices had positioned In-N-Out as a brand that could withstand the numerous industry hits and downturns to come—that is, as long as it stuck to its winning formula.

The industry had based its economic model in large part on profitability through aggressive growth. The strategy left many franchisees alienated, and franchising, which had helped to fuel the rapid expansion only a few years earlier, began to falter. As Robert McKay, the general manager of Taco Bell, told *Forbes* magazine, "Everyone wanted franchises in the mid-sixties. Then came the shakeout a few years later, and franchising no longer was the easy game it once was." To sustain growth, fast-food chains began penetrating foreign markets.

At the same time that a series of hits and downturns gripped the fast-food industry, beef prices were rising sharply. In order to maintain and reclaim customers and spike softening sales, the chains began to tinker with their classic menus. Over the next few years, nearly all of them began to introduce new items. McDonald's added its Quarter Pounder; in a bid to grab both the breakfast market and children's appetites, it also introduced the now-famous Egg McMuffin and later the Happy Meal. Wendy's began offering stuffed baked potatoes, salads, and chicken breast sandwiches. Over at the Carl's Jr. restaurants, Carl Karcher decided to launch a few fast-food firsts of his own. He began installing salad bars and he offered free refills on soft drinks. Although the addition of new items brought an initial surge in interest, it also introduced a new level of operational complexity with dubious long-term returns. Still, the strategy would be repeated over the years in an effort to shore up sagging sales. The race for the new American fast food marked a point of no return.

Soon the chains turned on one another. They undercut each other on price and tried to outdo each other with promotional giveaways, movie tie-ins, and a host of prize bonanzas. They erected colorful children's playgrounds. Even their ad campaigns began to reflect a showdown. Burger King launched its "Have It Your Way" campaign specifically to showcase that its customers could order their burgers individually. It was a public snub of the redoubtable McDonald's. Jack in the Box brashly goaded the industry leader with a series of ads in 1975 that ended with its employees shouting "Watch out, McDonald's! Watch out, McDonald's! Watch out, McDonald's!" The first salvo in what would be a recurrent battle labeled the "Burger Wars" had been fired.

In-N-Out Burger stayed largely removed from the war zone; this was not In-N-Out's fight. The family-owned burger chain didn't advertise, didn't undercut on price, didn't sacrifice quality, and didn't change the menu. At In-N-Out Burger, there were no sideshows.

As the atmosphere within the industry became increasingly competitive, its once picture-postcard Americana reputation soon began to sour. Those quirky individual stands and diners that had characterized the industry's start had given way to a formula of uniformity and homogeneity. Many chains faced growing opposition from neighborhoods across America. The resistance stemmed from what residents viewed as the compounded hazards of traffic, litter, crime, drugs, and delinquent teenagers; communities also objected to the chain restaurants on aesthetic grounds. In some areas, local ordinances forbade fast-food restaurants from opening near schools, churches, or even hospitals.

This backlash was directed at the industry as a whole, but its dominant player, McDonald's, took the brunt of the criticism. During this time, the residents of Woods Hole, Massachusetts, dotted the roads with homemade signs that read "No McDonalds Keep Woods Hole Franchise-free." In 1975, three community groups protested the efforts of one McDonald's franchiser to open in or near residential neighborhoods in Manhattan with the slogan: "We Deserve a Break Today. Stop McDonald's."

Many Americans were openly hostile to the practices of an industry that they felt trampled all over the landscape, the environment, the human body, and labor practices, all for a quick buck. Early critic of media manipulation Vance Packard, who wrote the 1957 best seller *The Hidden Persuaders*, complained sixteen years later to *Time* magazine about the ruinous effect that fast food was having on American values: "This is what our country is all about—blandness and standardization." Fast food's enormous popularity was seen by many as the starting point for the culture's decline.

Although seemingly impervious to many if not most of the poison darts targeting fast food, In-N-Out was not totally insulated from the larger standing of the industry, especially as the chain moved outside of its traditional home base in the San Gabriel Valley. The growing reputation of fast-food restaurants as teenage hangouts synonymous with unruliness, as well as crime magnets and totems of cultural imperialism, had inspired a strong disapproval across Southern California. "There was a lot of resistance to hamburger stands," contended Russell Blewett, who was elected Baldwin Park's mayor in 1972. "A lot of cities didn't like them. They thought they attracted a bad element."

In-N-Out didn't fall into neglect or become a hub of juvenile delinquency largely due to the efforts of Harry Snyder. Blewett, who called Harry "a Tough Dutch," first met Snyder when he was about ten years old. At the time, Harry gave Blewett a job sweeping up the In-N-Out parking lot on weekends. "He paid me in hamburgers and drinks," he recalled. Despite the thirty-year age gap, the two men became friends. "We forged a friendship. He liked me and I liked him." While other chains took to hiring security guards and implementing curfews for youngsters under eighteen to foil potential troublemakers, Harry "took things into his own hands," remembered Blewett. "He was very much about problem solving before it became a problem."

When Harry was interested in opening one of his drive-throughs in Rancho Cucamonga, a dry, dusty city on the edge of the Mojave Desert in San Bernardino where the Mojave Trail, the old Spanish Trail, the Santa Fe Trail, Route 66, and El Camino Real met thirty-nine miles east of Los Angeles, he called Blewett for help. In Blewett's version of the tale, at the time, Rancho Cucamonga was especially opposed to the idea of a new drive-through coming to town. "We had a meeting with the city's planning staff," he remembered. "And there was this planner just out of Cal Poly [College]. He was greener than grass. He said, 'I think your hamburger stands are ugly.'" With that Harry turned red and jumped to his feet and let loose with a string of expletives. After calming down, Harry turned to the young planner and demanded, "How many millions are you worth? How many successful businesses do you run?"

Recounting the tale, Blewett, who had at one time owned a successful floor coverings store in Baldwin Park, concluded: "I love that story. It's about city planners. People with no brains." In 1980, In-N-Out opened its first store in Rancho Cucamonga, followed by a second one thirteen years later.

CHAPTER 9

Guy and Rich Snyder were increasingly moving in different directions. Following high school, it had become evident that the Snyder brothers' diverging interests and personalities, which had first emerged when they were youngsters, had come into sharp relief as they grew into adults. As Rich was stepping deeper into the family business, his older brother, Guy, was moving further outside it.

A tough taskmaster of a father and a mother whose cheery outlook and blind faith held firmly to the notion that in the end things would turn out fine had bred two very different young men. From their appearance to their behavior, the Snyder brothers were like two sides of the same coin. Sandy-haired and casual regardless of the occasion, Guy favored jeans and T-shirts, while close-cropped brunet Rich always appeared to be cleanly and conservatively turned out. In a photograph of the two brothers flanking their mother sometime in the early 1970s, Guy is sporting a thick mustache and a confident smile; he is clad in a lively denim Western getup including a vest and a flowered, patterned shirt. A big silver belt buckle and silver tips on his collar completed the look. In his right hand, Guy is clasping a big cowboy hat—his left hand is cradling Esther's shoulder. Standing slightly behind his mother's left shoulder is a clean-shaven Rich wearing a timid smile. Wearing crisp, cream-colored trousers and a maroon pullover with a polo shirt, he looks as though he might be about to head off for a game of golf at the country club.

Like their parents, the Snyder brothers possessed an incredible sense of generosity. They both battled shyness and weight problems at different times and in various ways. Perhaps in an effort to counterbalance their relatively introverted natures, Guy and Rich both seemed to constantly surround themselves with a group of friends and associates—but it was the brothers' differences more than their common traits that most people noticed.

Guy could be moody, and his moods at times had a bad-tempered edge. He could be by turns sullen, quiet, irritable, confident, fun-loving, or incredibly kind. Where Guy was taciturn, Rich was gregarious and sociable. Although Rich could be a big kid himself, his fun was always tempered by his responsible nature. According to friends, Guy seemed most comfortable on the fringe. Ardently independent, he set his own boundaries.

As the radical 1960s passed into the hedonistic 1970s (an era later dubbed by writer Tom Wolfe "the Me Decade"), Rich held firmly to the concepts of religion, trust in government, and the nuclear family even as culture and society were shifting further and further away from those once-hallowed American ideals. As a young man, Rich found himself increasingly drawn to conservative politics. He became a great admirer of President Richard M. Nixon, who appealed to social conservatives, a moniker that Rich was rapidly taking on as his own. While all around him religion and traditional family structures were losing ground, Rich remained quite close to his parents, and rather than abandoning his spiritual beliefs, he was drawn ever closer to his Christian faith.

Early on, it became apparent that Rich was the obvious successor to Harry. A natural entrepreneur, he was ambitious, and friends noted that Rich shared Harry's populist touch. He was gifted with the kind of charismatic personality that moved people to follow his lead. At seventeen, Rich began taking care of In-N-Out's bookkeeping. When friends would ask him to go to a movie on a Saturday night he often declined: "I'm doing the books," he'd say.

Rich was exceptionally organized and driven; he regularly laid out a series of personal two-year, five-year, and ten-year plans (at the top of his personal ledger were marriage and family). Each year, Rich—accompanied by friends—went away for a weekend during which they worked out their goals and expectations together. "It's okay on goals to dream big," Rich once said, "because if you only get half, you've got a lot."

In fact, the Snyders' birth roles seemed to be reversed. Rich was more like the prototypical achieving firstborn son, while his older brother, Guy, was the unconventional nonconformist type often associated with a second or third child. In a way, each brother seemed to absorb the overriding social mores of the era in which he grew up. Rich embodied the 1950s of their childhood with its sense of order and simplicity, while Guy took on the antiestablishment rebellion that arrived during their teenage years in the 1960s. Not surprisingly, just as the culture of 1960s collided with the ethos of the 1950s, so too did Guy and Rich clash with each other. "They fought like cats and dogs," said one friend. As another friend put it, even as young adults, "Rich was the corporate type and Guy was the wild one."

Guy was restless; however, he was rarely at a loss to discover outlets for his considerable energy. He had a passion for speed—and some would say a recklessness—that he fed by racing his dragsters and riding motorcycles (and supplemented with alcohol and drugs). According to Wilbur Stites's wife, Kim (the pair first met while she was in high school and married in 1974), Guy liked to push limits. A local girl who worked at Snyder Distributing through college, Kim said, "He was always doing what he could get away with, without getting caught. Rich was always dependable, if you ever needed anything. But Guy, you had to find him first."

Then came the crash. In the mid-1970s, Guy was seriously injured when he lost control of his motorcycle, exploding over the sand dunes in the California desert at Glamis, on the Mexican border. As he hit the top of a dune with its razor-sharp drop-off, rather than kicking the bike away, he tried to ride it straight down. Instead, he was pinned beneath it. Wilbur Stites, who was racing with Guy on that trip, rushed him to the hospital. Many blamed the influence of drugs or alcohol (or both), but, Stites's wife, Kim, said that on that particular trip, Guy was "stone sober."

The accident had many persistent repercussions. For one, it left Guy severely injured and in chronic pain. He lost about 50 percent of the use of his right arm, and required numerous surgeries and therapy for his arm and back over the years. While hospitalized he was put on a morphine drip, which Harry insisted be removed as soon as possible. Whatever the degree of Guy's use of chemical substances before the crash, he developed an unassailable dependency on painkillers following the accident in an effort to help him cope with the constant pain. In fact, he never entirely recovered from the accident. As Paul Althouse exclaimed, "He was the nicest guy if you got to know him. But that was really the beginning of his downhill slide." From that point on, Guy's life was marked by cycles of depression, drugs, and many attempts to get clean and sober. Guy had begun his long skid into what friends referred to gently as "his troubles."

The situation seemed to exacerbate the already growing tensions between the Snyder brothers. However, under the strong hand of Harry, their complicated relationship appeared to remain stable. And in the short term, at least, it had little impact on the family business. In-N-Out Burger was not yet considered a major player in the fast-food industry. The private behavior of the Snyders was never aired publicly; that just wasn't Harry's or Esther's style. Unlike the showy displays of numerous entrepreneurs whose antics became fodder for rumor mills and newspaper columns, the family's personal business was rarely if ever linked publicly with the family's professional business.

Closer to home, even bigger changes were taking place; around 1974, Harry Snyder was diagnosed with lung cancer. The family was stunned. He had begun smoking cigarettes as a high school student in the 1930s when a pack of Chesterfields accompanied his regular game of rummy, but he had decided to quit in 1959. By then, however, the nearly thirty years of smoking had taken their toll.

Harry would not go down without a fight. He submitted to seemingly endless rounds of medical appointments, tests, and treatment. He ate specially prepared foods. Eventually, he endured chemotherapy; when his hair began falling out, leaving him bald, he took to wearing a wig. When his voice had dimmed to a mere whisper and he could hardly speak, Harry still managed to smile. Even as his body was succumbing to the advanced and aggressive cancer, Harry still hoped to find some kind of new or alternative treatment that might grant him a reprieve. At one point, he went down to Mexico to submit to Laetrile. An unconventional therapy not approved by the Federal Drug Administration, Laetrile, was based on using purified amygdaline, a chemical found in nuts and fruits like bitter almonds and apricots. Harry wasn't alone; during the 1970s, many American cancer patients, seeking any flicker of hope, flew to Mexico seeking new alternatives to rid them of their disease. And at one point, Harry thought he had his cancer beat.

It was Harry and Esther's intention to keep In-N-Out Burger a private, family-run business for succeeding generations of Snyders. Harry Snyder had very particular ideas about family and business; he had no interest in seeing his sweat equity disappear into some big company's idea of what In-N-Out Burger could be. On this point, he and Esther were in sync. His company was his family, but only his family had his blood. Esther was already into her mid-fifties; clearly, Harry's cancer hastened the need to set up a succession plan.

Successful family firms are often erected through the sheer will and force of a specific individual. The challenge for Harry was the challenge of all patriarchs—to pass on the company he built to someone who would be able to maintain its success without abandoning the unique culture that had made it a winning hand in the first place and to keep it in the family for successive generations.

Usually, that someone is the firstborn child. In-N-Out was both Guy and Rich's birthright. The drive-through was for a time practically their front yard, and they had each worked almost every position in the company, starting with picking up its trash. By having their boys work at an early age, the Snyders had hoped to instill in them a sense of responsibility and ownership. However, all things being equal, the brothers did not demonstrate equal promise to fill Harry's shoes. Perhaps it was because Harry was the classic self-made man, but he wasn't going to simply bestow leadership. His successor would have to earn it.

Guy felt strongly about the family business. But for some time, Guy had demonstrated that if he could be anywhere it would be behind the wheel of one of his dragsters chasing down a quarter-mile of asphalt. Furthermore, his recurrent troubles scarcely made him a strong candidate to take over the business; the crash in Mexico was hardly a ringing endorsement. Guy the so-called wild child had grown into something of a rebellious adult, exhibiting the kind of destructive behavior that prevented him from being named his father's heir and haunted him for the rest of his life. In some ways, his accident and its aftermath set the stage for the company's succession plan, which arrived sooner than anyone could have predicted.

Rich often accompanied Harry during his rounds of doctor's visits and treatments. Already close, the hours that father and son spent together seemed to bring them even closer. The pair spoke frequently about the business. Apparently, Harry used the time to relay to his son his thoughts and his goals concerning its operations, and his ideas for In-N-Out's future. Harry also reportedly recorded many of these discussions in a series of home movies. It was clear that Harry wanted In-N-Out Burger to continue, no matter what happened to him.

Even as Harry became gravely ill, he continued to make plans for In-N-Out. And the future as Harry saw it looked a lot like the past, with a limited expansion into the outlying communities beyond the San Gabriel Valley. In 1975, store number seventeen opened in Santa Ana, the chain's first drive-through in Orange County. About ten miles inland from the Pacific Ocean, Santa Ana was a growing city, one of the largest in Orange County. Orange County was the home of Disneyland and Knott's Berry Farm. A longtime Republican stronghold in California, it was also known for its famous beaches—one of them, Huntington, was famously dubbed "Surf City USA."

A year later in 1976, when McDonald's posted $3 billion in sales and opened restaurant number 4,000 in Montreal, In-N-Out opened its eighteenth store in Woodland Hills, a suburb in the southwest San Fernando Valley near the 101 Freeway on Ventura Boulevard. It was another ace spot for the chain. Originally part of the El Camino Real (the Royal Highway) that linked twenty-one Spanish missions, Ventura Boulevard was also one of the main east-west thoroughfares in the Valley and was the original U.S. Route 101 before the freeway was built. The Woodland Hills location, like most In-N-Out Burgers, was positioned to take advantage of the maximum amount of traffic. Heavily traveled, the 101 Freeway in Southern California followed the Pacific Coast to the beaches running down to Hollywood and up to San Francisco and onto Oregon in the north. Additionally, the busy boulevard was flanked with numerous small shops and businesses. This was the third In-N-Out to open in the San Fernando Valley in five years, and customers from all points of the northwest valley made special trips just to eat at the new location.

The Woodland Hills drive-through was the last In-N-Out Burger opening that Harry Snyder oversaw. He died on December 14, 1976; he was sixty-three years old. Harry's funeral was held three days later at 10:00 a.m. at the white colonial-style Church of Our Heritage on the sprawling grounds of the Forest Lawn Memorial Park and Mortuary in Covina Hills. It was a fitting final resting place for Harry as well as a slightly odd one.

Like In-N-Out, Forest Lawn, a chain of cemetery parks, was a uniquely Southern California phenomenon. Just as Harry Snyder remade American dining with his drive-through burger chain, Hubert Eaton, the founder of Forest Lawn, had revolutionized the funeral industry. In 1917, the somewhat eccentric Missouri-born businessman created a chain of memorial parks across the Southland that broke with what he believed was the usual dreary and depressing cemetery setting. Featuring scrupulously manicured grounds lined with trees, fountains, music drifting out of speakers hidden in rose bushes, and grave markers flush to the ground to give them more of a park feel,

Forest Lawn parks were built to be as "unlike other cemeteries as sunshine is unlike darkness, as Eternal Life is unlike Death." *

Eaton commissioned hundreds of statues and artwork including reproductions of some of the world's most famous works of religious art, such as thirteen mosaic scenes of Michelangelo's Sistine Chapel (as well as kitschy original renditions: a 172-by-35-foot mosaic depicting twenty-six famous scenes from the earthly life of Jesus) to decorate the Forest Lawn chain. Eaton introduced a "pre-need" program that allowed people to see to their own funeral arrangements before they died. Soon, Eaton lured couples to marry in the chapels on the cemetery grounds. It was at Forest Lawn's Wee Kirk O' the Heather Church that Ronald Reagan married his first wife, actress Jane Wyman, on January 26, 1940.

On a warm winter morning, Harry's funeral mourners filed into the 120-seat church, a replica of St. George's Church in Fredericksburg, Virginia. The chapel was too small to accommodate the hundreds of mourners who had come to pay their last respects to Harry Snyder, and many stood outside. "He really touched a lot of folks," said Russell Blewett, who attended. Numerous associates and community residents arrived. It seemed that everybody that he ever had contact with arrived to say good-bye. The Snyders' good friends Carl and Margaret Karcher, who had stayed close through the years since they first met and offered Harry advice on running a fast-food shop, paid their last respects, too. Three years later, the Snyders' foster son, Wilber Stites, was killed in a car accident. He was buried near Harry at Forest Lawn.

When Harry died, the question of what would happen next to the popular little burger hut with the red-striped awning had already been de-

---

* In 1966, ten years before Harry died, Walt Disney's cremated remains were interred at Forest Lawn's Glendale memorial park—not (as widely rumored) cryogenically preserved in order to bring him back to life one day in the future when technological advances would be able to thaw his body and cure his disease.

cided. At fifty-six, Esther Snyder was now a widow with few financial concerns. She might have easily jettisoned the chain and taken up a life of early retirement; instead, she agreed to carry on. In fact, Esther insisted upon doing so. She simply cared too much about the welfare of her associates to abandon In-N-Out. They had shown such loyalty to the Snyders that she couldn't imagine pulling the rug out from under them now. Besides, she was still devoted to Harry, and In-N-Out was his legacy. "I think she just didn't know what else to do," remarked old family friend Valerie Althouse. "Esther never sold it because of Harry. It was a family business. Just about everyone working there had been there since the 1960s. They had become like family."

For a short time, there was some thought of Rich possibly doing something else. This was, however, mostly fueled by a brief spasm of personal doubt. Although he had not attended college, Rich had learned the business by working there. In short, he feared that he might not have what it took to run In-N-Out Burger. On a few occasions, Rich delicately broached the idea of selling the company with his father before he died. But Harry dismissed his son's misgivings; he'd put his arm around his son and tell him, "You can do it, Rich."

Seven months shy of his twenty-fifth birthday, Rich Snyder was named president of In-N-Out Burger, and whatever reservations he had, he had quickly pushed them aside. Esther was given the official title of secretary-treasurer (she also retained controlling interest in the company). Guy was named In-N-Out's executive vice president. But Rich was also named the trustee for his father's estate and trust instrument. For all intents and purposes, Rich was given the keys to the family's growing burger kingdom. Guy Snyder was passed over in favor of his brother.

Practically speaking, the decision to hand the company over to Rich was both sensible and prescient. In retrospect, it was the plot point that turned In-N-Out from a local eatery into one of the largest family-owned restaurant chains in the country, competing head-on with the fast-food giants. Of course, as with most plot points, the narrative doesn't turn without a certain dramatic tension—and that tension was derived in no small part from the relationship between the Snyder brothers. The upending of primogeniture created a conflict between the two brothers whose relationship was already exacerbated by personality clashes. It had the emotional intensity of the biblical story of Jacob and Esau without, of course, the bloodlust or deceit. As his longtime friend explained, "Guy never said outright, 'I really got screwed over by dad' or anything. Guy knew he had a problem with the drugs and I don't think he ever questioned why he was passed over. But to tell you the truth, he was hurt nonetheless."

Outside In-N-Out's fortified walls, the wholesome image of the company remained intact; behind them, however, there was a small fissure, like a hairline crack in a crystal vase.

CHAPTER 10

Harry's confidence in his youngest son was not misplaced. Whatever initial misgivings Rich may have had, they were nowhere in evidence when he seized the reins of the family business. In-N-Out's new president took up his position filled with the kind of determination and enthusiasm that characterized him both inside the fast-food industry and outside of it. Arriving early each morning at the chain's nondescript Baldwin Park offices (where Snyder Distributing was housed across from store Number One), Rich cut a commanding if not boyish figure, aided perhaps by his large girth. Not yet twenty-five years old, he was almost always dressed in a suit and tie, and just as often a broad smile. Rich possessed the kind of clean-shaven face that seemed to magnify his already youthful appearance.

It was not, however, an entirely smooth transition. Very quickly, Rich discovered that some of the associates that he had considered to be his friends did not always have his best interests at heart. And at twenty-four, he realized that he had a lot to learn. When describing this period of adversity and hard knocks he later said that "the bumps along way just help to develop character." However, from that point forward, Rich never again considered selling In-N-Out. The very idea had become a taboo subject.

By the time that Rich took over In-N-Out Burger, the fast-food landscape was clearly different from the one that his parents had helped establish. For starters, there was a strong public perception that hamburgers were mostly junk food; the industry was now stuck with the negative reputation it had earned as a business of cheap, low-quality food prepared by disposable, underpaid workers. Still, despite its bad press, the fast-food world continued to prosper. In 1976, the United States was celebrating its bicentennial and the fast-food industry was ringing up $16.3 billion in sales. As it continued to grow at home, the business had gone global.

For Rich, a man who believed in the hamburger like he did the American flag, the public's gloomy perception of his business was particularly irksome. It was an opinion Rich didn't share in the least—he had nothing but respect for the hamburger business, especially In-N-Out.

Whether Harry expected Rich to follow in his cautious footsteps is unclear, but certainly his son had some ideas of his own. Unlike many heirs who seek to make their mark on the family firm by completely remaking it, when Rich took over In-N-Out Burger he saw the wisdom in maintaining his parents' 1948 formula. To his credit, Rich wasn't tempted to tinker with In-N-Out's limited set of offerings even as his competitors had, for some time, rapidly expanded their own. In the chain's history to date, it had only added one new product to its menu—and that was the soft drink 7-Up. "It's hard enough to sell burgers, fries, and drinks, right," was how Rich explained his reasoning. "And when you start adding things it gets worse."

In-N-Out's limited scope and narrow focus also meant that the chain didn't have to continuously spend money on new equipment needed to prepare and cook new menu items. By the same token, it wasn't necessary to repeatedly train its associates to learn how to ready those new offerings. And without shareholders peering over their shoulders, the Snyders were free to spend their money where they wanted—and that was on maintaining high standards.

That's not to say that Rich Snyder did not have his own dream for the future of In-N-Out Burger; in fact, he had big plans for the chain. His vision mirrored his father's with one exception. If Harry was a man who was satisfied with his limits, his son was a man who saw greater potential. By the time Harry died, he was happy with the small clutch of stores that he had prudently rolled out across the Southland. Those who knew him said he was reluctant to push the chain much beyond that. The eighteen In-N-Out units ringing the San Gabriel and San Fernando valleys as well as Orange County were a sufficient measure of success, to his way of thinking.

His son, however, had his own measure of success. The chain that Rich had inherited was a cherished brand with a devoted following. In his mind, Rich firmly believed that he could maintain his parents' model of simplicity and quality while greatly expanding In-N-Out Burger. Initially, he decided to venture farther south of Los Angeles County, push farther into the fast-growing Orange County, and move into new territories, starting with Riverside and San Bernardino counties. "Rich was very unique," explained his cousin Bob Meserve. "He had a lot of his dad's qualities. But Rich was shrewder as a businessperson. Harry was old school and Rich was new school. Rich had a vision. He knew what he what he wanted to accomplish."

What Rich was not, however, was a college graduate. Despite his considerable natural talents, it was something that seemed to leave him slightly insecure. His dyslexia, diagnosed only when he was an adult, amplified the feeling. Close observers noted that Rich lacked the polish of many executives, but what he lacked in luster he more than made up for with his genuine ability to connect with people. When he gave a speech, more often than not, it came straight from his heart.

In any event, Rich strove to continually improve his management abilities, taking advantage of opportunities to expand his own education at every turn. He cobbled together his own kind of degree, attending leadership seminars and classes and seeking out mentors. Rich became active in a number of business organizations and peer groups where he met and mingled with other Southern California business executives. He belonged to the networking group The Executive Committee (TEC), made up of twelve company presidents from various businesses and industries. He also joined the Young Presidents' Organization, a worldwide peer group exclusively for CEOs and presidents of companies who were under the age of forty-five.

The affiliations gave Rich the opportunity to mix with other self-starters as well as second- and third-generation businessmen who had also inherited their family businesses. They provided a kind of fraternity as well as forums in which he could candidly and safely discuss issues, solve problems, and exchange a range of ideas on everything from marketing to taxes to human resources. Within this circle, Rich was universally well regarded for his energy, enthusiasm, creativity, and his obvious gifts as a leader.

It was during this time that Rich met Jack Williams. Originally from Clovis, New Mexico, Williams and his wife, Linda, were longtime restaurant business veterans. The couple had come to California in 1954, when Williams was part of the U.S. Marine Corps and was stationed in San Diego. In 1957, the couple began working in the food business, and in 1969 they opened their first restaurant, a Sizzler Family Steakhouse. First established in 1958, Sizzler was a pioneer in the fast casual dining segment and became known for its reasonably priced steak dinners and all-you-can-eat salad bar. Williams went on to become one of its largest franchisees.

The two men had frequently run into each other in industry circles. But when Rich and Williams sat on the board of the California Restaurant Association together, they became fast friends. Williams's initial impression, he recalled, was that Rich "was blessed with a tremendous amount of common sense and a lot of business ability," and he possessed a real determination to improve and learn.

Although Williams was eighteen years Rich's senior, the two men spent a lot of time together problem solving, attending training seminars, and generally using each other as professional sounding boards. The pair served on each other's advisory boards; they shared an interest in the operational details at every level, frequently eating at each other's restaurants, reporting back any problems. "Many times I told him about little things," remembered Williams. "They wouldn't be noticed by the public."

Williams and his wife owned a ranch down in Temecula in Riverside County, a town known for its floating pageant of hot air balloon festivals, and Rich came down at least once a month, often arriving by helicopter. There they rode horses together. In fact, after Rich bought his own copper red Tennessee Walker (whom he named Ernie), he boarded him for a time at the Williamses' ranch. In Temecula, the two friends fell into a routine of packing a lunch of peanut butter and jelly sandwiches and riding the property out to the nearby campgrounds while talking about businesses matters. "We'd ride up there and go sit under the oak tree and work on our employee handbooks," Williams recalled. "He had a real innate sense."

Less than two years after he became In-N-Out's president, Rich's management skills were put to the test. At twenty-six, he faced the first major challenge to his abilities as a leader and the future of In-N-Out Burger.

At about 7:00 p.m. on August 16, 1978, a fire broke out at the chain's Baldwin Park headquarters. Sixty firefighters from eleven engine companies rushed to 13502 E. Virginia Avenue to battle the flames as they ripped through In-N-Out's warehouse, offices, and meat department. The blaze had caused the sky to turn black, filling it with giant plumes of smoke that could be seen as far away as the cities of Duarte, six miles to the north of Baldwin Park, and Hacienda Heights, eleven miles to the south.

When the smoke had cleared, the facility was completely gutted. The roof had collapsed, and only the burnt walls of the hollowed-out concrete warehouse remained. Fortunately, the facility was closed at the time and nobody was inside. The following day, the *San Gabriel*

Valley Tribune reported that the blaze had caused an estimated $750,000 in damages.

The fire was potentially devastating for In-N-Out Burger, and not just in monetary terms. The warehouse, built before sprinklers were required, was the heart of In-N-Out's operations. The fire didn't just pose a threat to Rich's immediate plans to expand; it jeopardized In-N-Out's ability to continue to operate, period. But Rich was determined to get In-N-Out back up and running immediately.

Resuming the chain's operations was a scramble of logistics and perseverance. Rich and Esther bravely and quickly organized a game plan. Carrying the Snyders through this rough period was the amazing show of support by the chain's associates.

While the associates did their part, even more important, Rich was able to rely on In-N-Out's close ties with its suppliers. Like his father before him, Rich continued to stick by the company's promises to pay full price for the highest quality ingredients. When prices plunged or spiked, or there were shortages due to weather or other events, In-N-Out always absorbed the cost. As long as the quality remained exceptional, he did not look for cheaper suppliers. It was part of the Snyders' business practice to take care of their purveyors as they did their customers and associates. After the fire, when In-N-Out was unable to deliver supplies to the chain, a group of In-N-Out's longtime suppliers delivered their goods to individual stores until the warehouse was rebuilt. In-N-Out Burger had always taken care of its suppliers, and, following the warehouse fire, this time the suppliers took care of the burger chain.

Esther took charge of the chain's administrative duties, converting a portion of the first floor of her San Dimas house into In-N-Out's temporary offices. With eighteen stores, In-N-Out Burger was still a relatively small operation, and Esther brought a handful of "the girls," as she called them, from accounting to work with her in her home. Rich found a rental facility nearby that he turned into the chain's provisional warehouse and distribution center. And he worked out of the small warehouse building on the edge of store Number One that still housed Snyder Distributing.

While the Snyders made plans to rebuild their operational and administrative facilities on East Virginia Avenue, they ran the company in this manner for nearly two years. To Rich's considerable credit, each one of the chain's eighteen stores remained open and running with hardly a hiccup. In fact, while still operating the company out of temporary facilities, Rich demonstrated that the fire would not hamper his plans to move In-N-Out Burger forward. By 1979 he had added three new stores, including the first in San Bernardino Country.

It had become clear that as In-N-Out moved ahead with its expansion it needed to construct its newer stands to accommodate larger crowds. It had outgrown its signature format: a simple 250-square-foot open kitchen flanked by twin drive-through lanes, with a walk up window (and nearby five hundred-square-foot storage building) on a one-acre plot. While that prototype helped spawn a host of imitators, the Snyders were running up against two crucial issues: the growing cost and dearth of available one-acre plots and the chain's own unbridled popularity.

Bottlenecked drive-through lanes spilling into passing street traffic had become almost as synonymous with In-N-Out as its boomerang arrow. Increasingly, municipal zoning officials began to think twice about signing off on permits for In-N-Out's twin-lane format, causing numerous delays for new openings. Intent on expansion, Rich found an alternative solution to increase capacity and keep city officials placated: build a new store prototype with indoor dining and a single drive-through lane. The Ontario store was the first designed with the new prototype. As Rich later told *Nation's Restaurant News*, "I think double drive-throughs are great; we love to build them. But most cities [resist issuing permits to high-volume drive-through operators], at least if their name is In-N-Out." In some ways, the shift marked the end of the Harry Snyder double drive-through era.

Meanwhile, Guy Snyder was pursuing a life mostly outside the family business. Officially, he was In-N-Out's executive vice president, but it had been obvious that it was Rich who was running the company. He was earning a salary that was reported to be close to seven figures although he was not an active presence in the company. Guy's attentions were focused elsewhere.

Much of Guy's time was spent on the drag circuit. Although his motorcycle accident had left him with only about 50 percent mobility in his right arm, in frequent pain, and in need of subsequent back and arm surgeries, Guy continued to find a way to race. The devastating injury made it virtually impossible for Guy to shift his four-speed. Even a task as banal as zipping up his own fire suit at times proved difficult, and he could be seen struggling, his right arm limp at his side, as he closed up his jacket. The injury made it difficult for him to obtain his racing license, and so Guy had an automatic transmission built for his cars and raced sportsman class instead.

Along with his brother, Guy had amassed a significant collection of rare and classic hot rods. It was an expensive hobby that dated back to his teenaged days at the Dale. The collection included a number of jewels such as a 1963 Corvette Split Window Coupe and a 1965 Shelby Cobra 427. Guy was said to have paid about $500,000 for the Cobra that was featured in the 1966 Elvis Presley film *Spinout*.

Guy converted a one-hundred-by-seventy-five-foot storage space within the East Virginia Avenue complex into a museum-quality hangar for the cars with a black-and-white checkered marble flooring fit for an Italian palazzo. The hangar was far enough east of the original warehouse that it was spared from the 1978 fire. Just outside his hangar, Guy built a miniature drag strip about one-eighth of a mile long along with an NHRA regulation burn-out pad and timing light where he could test the cars' performance.

In 1979, while staying at one of the family's vacation homes in the tony surfers' haven of Hermosa Beach, Guy met a woman named Lynda Lou Perkins (*née* Wilson). At the time, she was working in a T-shirt shop and the two began dating. Lynda was seven years older than Guy and had two daughters, Traci (thirteen) and Terri (eleven), from a previous marriage. According to intimates, it was Lynda who pursued the eligible bachelor.

ARMY GREEN: A perforated eardrum left Harry Snyder, circa 1942, stateside and serving largely behind a desk during World War II. Among his duties, Harry processed B-52s at Hamilton Field in Novato, California. On the side, Harry worked at the Sausalito Shipyards for extra cash. (*Rich Snyder Family Collection*)

NAVY BLUE: Esther enlisted in the Navy WAVES (Women Accepted for Voluntary Emergency Service) in 1943. Established to fill the vacancies created by the thousands of men sent to the battlefront, WAVES were not eligible for combat duty; following boot camp, Esther was stationed at the San Diego Naval Hospital, where she earned the rank of pharmacist's mate second class. (*U.S. Navy Memorial, Washington D.C.*)

BOATING IN SEATTLE: Harry and Esther met in 1947 at Fort Lawton, where she was the restaurant's manager and he sold sandwiches. After the war, Harry came up with the idea for a new kind of restaurant. (*Rich Snyder Family Collection*)

HALCYON DAYS: In the fields and orchards surrounding the Snyders' San Dimas house, Guy and Rich (circa late 1950s) played army and Guy displayed his early love of cars. (*Rich Snyder Family Collection*)

NO DELAY: Harry's two-way speaker box inspired the company's name—customers drove *in* to order and then drove *out* without ever having to leave their cars. The original (1948) sign is now pitched at the company's Baldwin Park headquarters. (*Duke Sherman*)

BURGER U: In 1984, Rich Snyder built the chain's first university on the site of Harry and Esther's old Baldwin Park house. The university's purpose was to help create and train a steady pool of management talent to populate the expanding chain. The current university was built in 2004. (*Duke Sherman*)

ROADSIDE BILLBOARD: The famous In-N-Out yellow arrow and analog clock tower still stands on the north side of the I-10 freeway. In 1954, the path of the I-10's expansion in Baldwin Park cut through store Number One and the Snyders tore down the original, rebuilding a new Number One a short distance away. Harry designed the new shop as the now famous double drive-through. In 2004, In-N-Out shuttered the original and built the third of its "Number Ones" on the opposite side of the freeway. (*Duke Sherman*)

THE WEDDING PARTY: Rich and Christina Snyder (center) married on May 2, 1992, in Maui. Almost eighteen months later Rich, Phil West (far left), and Jack Sims (third from left) were killed in a plane crash. (*John L. Blom Custom Photography*)

IN-N-OUT BURGER FAMILY PICNIC: At the annual company outing in 1997, Guy Snyder and his second wife, Kathy Touché, along with her children, Aaron, Andy, and Emily. (*Kathy Touché*)

THE PRESIDENT AND SECRETARY-TREASURER CIRCA LATE 1980S: Despite their great success, Rich and his mother, Esther, loved attending new store openings and visiting with associates. (*Rich Snyder Family Collection*)

IN-N-OUT BURGER'S SOLE HEIRESS: Lynsi Snyder Martinez has been largely kept out of the spotlight and rarely photographed publicly. As a toddler in the arms of her Uncle Rich (circa 1984), she's already aware of her inheritance, holding an In-N-Out soda cup. (*Rich Snyder Family Collection*)

MICHELIN-STARRED FAN:
The French Laundry's renowned chef Thomas Keller (left) was photographed in the Napa Valley In-N-Out store for the April 2006 issue of *Food & Wine* magazine. A longtime (and vocal) fan of the chain, Keller announced he was opening his own burger restaurant, Burgers and Half-Bottles, an homage of sorts to In-N-Out. (*Thomas Heinser*)

BREAKOUT DESIGN: The Westwood In-N-Out, store 119, was designed by famed Los Angeles architect Stephen Kanner in the pop modern style in 1997. Kanner's hopes to leverage his award-winning design to re-brand the chain going forward were dashed when he was told "The burgers should be the star," not the store. (*Kanner Architects*)

IN-N-OUT AT THE OSCARS: Since 2001, *Vanity Fair* editor Graydon Carter has hired one of the chain's cookout trailers for his annual party at Morton's following the Academy Awards. Wildly popular, the iconic burgers have become an integral part of the exclusive fete. At the 2006 party, alternative rocker Beck grabs a cheeseburger. (*Eric Charbonneau/ Wireimages Getty Images*)

THE QUEEN: Celebrated actress Helen Mirren in her custom-made Christian Lacroix gown was famously photographed tucking into an In-N-Out burger following her best actress win at *Vanity Fair*'s 2007 party. (*Eric Charbonneau/Wireimages Getty Images*)

After courting for a year and a half the pair married in 1981. However, Rich insisted that his brother have a prenuptial agreement signed before the wedding could take place. Just two days before the wedding, Kim Stites ran into Lynda, who appeared irritated by the prospect. "She wasn't happy about the prenuptial," Stites recalled. Nevertheless, Lynda apparently signed it, as the event moved forward.

Their wedding on Valentine's Day was a lavish affair held at the Crystal Cathedral in Garden Grove. The Philip Johnson–designed church took its name from the more than ten thousand rectangular panes of silver glass that housed the cathedral. The new Mr. and Mrs. Snyder began their marriage in high style; a helicopter flew the couple to their reception aboard the *Queen Mary*, the luxury liner that had once ferried Winston Churchill, Marlene Dietrich, Clark Gable, and Fred Astaire across the Atlantic.

By all accounts, Guy transitioned easily from a thirty-year-old bachelor to a husband with a ready-made family, quickly embracing his two new stepdaughters. Said one relative, "He loved those girls like his own." The couple lived in a $600,000 estate home in Glendora, nestled in the bluffs at the foot of the San Gabriel Mountains. The couple divided their time between Glendora and their 170-acre ranch in the picturesque mountain wilderness of Shasta County at the northern ridge of the Sacramento Valley. Up at the Flying Dutchman Ranch (named in homage to his father Harry's ancestry), Guy liked to ride his Caterpillar tractor across the property, play video games, engage in paintball matches, and shoot off pistols. In the summers, he went fishing in Alaska or hunting in Montana. Lynda began raising llamas.

Guy was not completely absent from In-N-Out; he was involved to a degree with the warehouse and meat department. Along with Lynda, he attended many of the company's events and activities—but there was no question that it was Rich who was in charge at In-N-Out Burger.

On May 5, 1982, Lynda gave birth to the couple's first child—a daughter, Lynsi Lavella Snyder. She was Guy's only biological child, and he doted on his new daughter. It appeared that, despite his recurrent troubles, Guy had found a measure of stability in his life.

CHAPTER 11

Rich's tastes ran contrary to those of his parochial parents. He had no problem spending extravagantly and often turned the most basic of events into glittering, over-the-top affairs. Straightlaced in practically every other area of his life, his prolific spending stemmed from his belief that In-N-Out was not just a fast-food joint—and Rich's predilection for the showy was his way of broadcasting this belief.

In January 1979, Rich hired a local architect to work with him on designing an entirely new headquarters and warehouse on the original property on East Virginia Avenue. In rebuilding the facility, Rich decided to erect a structure that would incorporate the practicalities needed for the growing company he envisioned and at the same time embody Rich's lofty ambitions. When it was finished, it resembled one of the mansions of the captains of industry rather than the offices of a modest Depression-era couple who founded a postwar burger joint.

Construction crews broke ground in Baldwin Park in 1979, and by the time In-N-Out Burger's sprawling nine-acre complex was finished almost two years later, the chain had expanded to twenty-four shops. On December 1, 1981, In-N-Out's corporate office personnel moved in, signaling a new era for the small burger chain.

It was a time of dynamic changes. Just eleven months earlier, former movie star and governor of California Ronald Reagan was inaugurated as the fortieth president of the United States. The Republican administration marked a new era of conservative politics and economics. The national news was focused on the new president's supply-side economic policy (called "Reaganomics"); the end of détente with the Soviet Union; and the start of a massive military buildup.

On the cultural front, a new nighttime soap opera revolving around a fictional wealthy oil family from Denver called *Dynasty* made its debut. MTV, the twenty-four-hour music video channel launched on cable. And a shy, twenty-year-old kindergarten teacher named Lady Diana Spencer became an instant global phenomenon when she married Britain's Prince Charles on July 29, 1981. Their wedding attracted 750 million television viewers in sixty-one countries.

The timing was perfect. The new In-N-Out headquarters' ribbon-cutting ceremony took place just as the San Gabriel Valley was on the cusp of experiencing tremendous economic growth. Baldwin Park's chamber of commerce began actively recruiting new businesses and redevelopment projects. A newly built two-bedroom house with two bathrooms cost about $80,000—a tenfold increase from the time when Esther and Harry first arrived.

Located among gritty industrial factories and lumberyards, In-N-Out Burger's new complex ran along both sides of East Virginia Avenue, encircled by a sentry of long-necked palm trees, securely locked behind gates. The headquarters' showpiece was its two-story villa, modeled after one of the many grand Spanish mission-style revival villas that had once graced the San Gabriel Valley. It featured stucco walls, terra-cotta roofing, arched windows and doorways, hand-carved woodwork, vintage tiled ceilings, and wrought-iron ornamentation. A custom-made stone fountain was flown in from Guadalajara, Mexico. Rising above the villa was a custom-made cupola with a twelve-foot diameter and a five-foot drop. Inside the corporate offices, a large portrait of Harry Snyder was hung.

Instead of asphalt, Rich chose to lay down costly poured concrete. Custom landscaping brightened up the islands of concrete. Manicured lawns were planted, as were beds of flowers, and dense hedges were trimmed to spell out in block letters "IN-N-OUT." In a nod to the chain's origins, the original red and white neon sign from the first stand that read "IN-N-OUT HAMBURGERS NO DELAY" was posted on a patch of grass outside of the villa not far from the entrance gate.

Inside, the first floor was made up of several departments and rooms including the automotive department that handled all of the administrative aspects of operating the chain's growing fleet of eighteen-wheelers, maintained at the chain's truck depot that flanked the headquarters. In the "Liberty Room," used for managers' meetings and social functions, there was an antique oak bar, built by the Brunswick Company (the manufacturer of the famed billiard tables) in 1887. According to a detailed company inventory, the ground-floor accounting department was built with a vault that held the company's "pertinent records and in-house computer," where the temperature was maintained at a constant sixty-five to sixty-eight degrees.

A custom-made staircase with handmade railing and turnings bridged the second-floor executive offices, separated by antique doors imported from Spain. The suite's ceilings were made of lighted stained glass. Rich had the largest of the offices. A voracious reader of politics, his workspace was filled with shelves lined with books and decorated with an eclectic assortment of mementos such as his coin collection, photographs of various Republican apparatchiks, and Rich's prized 1944 Wurlitzer jukebox. Although he was rarely there, Guy had an office on the second floor, too. Smaller then Rich's, Guy's office was filled with mounted fish and game as well as drag racing memorabilia. Sumptuous and decidedly feminine, Esther Snyder's office was decorated with a suite of antique Louis XIV furniture more in keeping with Rich's idea of the kind of office that Esther should have rather than what she herself might have designed. In fact, she was hardly ever found behind her second-floor desk. Most of the time, Esther was downstairs with her sleeves rolled up, working among the cubicles in accounting.

When it came to rebuilding the warehouse, meat department, and other ancillary facilities, Rich had very specific ideas that built on Harry's original concept of total quality control. The function of the new warehouse remained the same (it would still receive, store, and ship food supplies to the stores), but it performed on a larger scale. For instance, In-N-Out retained a potato buyer whose job it was to select the proper quality potatoes for the chain's fries.

However, if anything paid homage to the ideals of Harry Snyder and In-N-Out's unwavering commitment to freshness and quality control, it was the new warehouse and meat department. Rich exhaustively researched slaughterhouses and meat processing plants around the country for best practices before constructing the new processing center (crowned with a Spanish mission-style bell tower) that operated seven days a week.

Inside, specially selected cow and steer chucks arrived at the antiseptically clean commissary. The chain proudly proclaimed that it paid "a premium to purchase fresh, high-quality beef chucks." And it required all of its beef suppliers to sign a purchase specification agreement prohibiting them from using "downer" (non-ambulatory) cattle. To better enforce In-N-Out's quality standards, each chuck was inspected before being accepted.* After In-N-Out's inspection, a team of skilled butchers boned and removed the meat. A blend of the meat was then put through a double-grinder—with the first mill breaking it down and the second readying the beef for five special machines that molded the beef into patties at a rate of twelve hundred patties per minute. Boxed and loaded into In-N-Out's refrigerated trucks, the finished patties were then sent off to the drive-throughs. In spring 1984, the *San Gabriel Valley Tribune* reported that the family-owned chain was selling more than 14 million burgers each year.

Five years after the rebuild, having grown to thirty-two stores (with thirteen more set to open), Rich needed to enlarge and update

---

* On January 30, 2008, In-N-Out released a statement that it had ended its relationship with the California meat supplier Hallmark/Westland after inspectors found that the company used downer cows, and the USDA initiated a huge recall of the company's meat. In-N-Out assured its customers that no meat from Hallmark/Westland was in the company's system and that it never purchased processed patties or ground beef from any suppliers.

the warehouse again. Roughly two dozen local business executives and a handful of Baldwin Park city council members attended the ground breaking of the new $3 million facility. The *San Gabriel Valley Tribune* featured the event in its February 11, 1986, edition with a large photograph showing Rich Snyder and four city councilmen (all in suits, ties, and hard hats) on-site.

By 1990, Rich had once again doubled the size of the chain; there were now sixty-four restaurants. In-N-Out's growing critical mass (and continued profitability and popularity) spurred analysts to estimate the private company's sales. *Nation's Restaurant News* reported that In-N-Out was generating roughly $60 million in sales annually. However, Technomic Inc., a Chicago-based restaurant consultancy that published a series of industry surveys, put the figure closer to $73 million.

From the very beginning, it had been the Snyders' policy not to divulge the company's numbers. They ignored all requests to do so, regularly frustrating the financial community. As Technomic president Ron Paul put it, "They are very different. They don't allow us to interview them or provide any data. We try at least once a year. They just don't respond."

At the start of the new decade, once again Rich renovated the warehouse to keep up with the growing demand, this time expanding it to seventy-five thousand square feet. That In-N-Out was able to maintain its fresh, quality, flavorful burger even as it scaled up was a distinction of which Esther was especially proud. As she once explained (in remarks published by the Baldwin Park Historical Society not long after the new headquarters were built), "The aims set forth by Harry Snyder since the founding of the company are still our chief endeavor—'Quality, Cleanliness, and Service.'"

Before long, investors had begun to take notice of the popular little Baldwin Park–based burger chain. They began sniffing around— only to be politely but firmly rebuffed. Voicing the desires of many, David Geraty, at that time the managing director of the Minneapolis investment banking firm Dain Rauscher Wessels, once exclaimed, "Every investment banker in the country would love to take them public." As early as 1986, Rich remarked that he had to deny the IPO rumor "at least twenty-five times a week." Although flattered and certainly aware of the potential financial windfall of such a move, Rich showed little interest in the attention given by excitable investment bankers. "In-N-Out is a great vehicle to do something like that," he once confessed. "But my feeling is that I would be prostituting what my parents made by doing that. There is money to be made by doing those things, but you lose something, and I don't want to lose what I was raised with all my life."

Similarly, In-N-Out routinely rejected frequent requests to franchise (reportedly thirty to thirty-five a week). After a time, they didn't even entertain the inquiries. The family saw franchising as a surefire path to losing quality control. "Franchising," Rich stated firmly, "was simply not going to happen."

In-N-Out's new headquarters became something akin to Baldwin Park's mansion on the hill. On March 16, 1982, approximately three months after moving in, the Snyders erected a large tent outside of their corporate offices and threw a dinner and dancing party for nine hundred guests. Despite the light rain, inside the tent the party resembled a summer night. Fairy lights were strung overhead, while rich carpets were laid out over the concrete. It was a huge event among locals. The party warranted inclusion in the historical society's inaugural issue of *The Heritage of Baldwin Park* newsletter.

The Snyders held numerous invitation-only parties and mixers for the city's chamber of commerce and a few other organizations on its premises. Semi-regular events such as "Business After Hours" were hosted by Rich and his mother, Esther, and usually included door prizes and food donated by the family. "Come join us for an evening of fun, fellowship, and great-tasting In-N-Out burgers!" announced a typical invitation.

On special occasions, the Snyders opened up their headquarters to select members of community groups, city officials, and area residents. They were given tours of the facilities, including a chance to explore the warehouse and commissary that never failed to impress visitors. So coveted were the invitations that they were received not unlike winning the golden ticket to enter Willie Wonka's chocolate factory. Eighty-five-year-old Bobbie Lightfoot, a longtime Baldwin Park resident, remembered the time she was invited on a tour in 1982. "They arranged for some of us real old-timers to go in vans," she said animatedly. "What a beautiful building. There was a fancy fountain and mahogany wood. Boy, did Rich make that a showplace." During the visit, Lightfoot pulled Esther aside and told her that "my grandson just loves In-N-Out, but he live[s] in Washington and [is] waiting for [you] to go there. She laughed and said, 'We're getting there.'"

Manuel Lozano (who became Baldwin Park's mayor in 1999) was a young city council member when he received his one and only invitation to tour In-N-Out's headquarters. "It was a rare treat, because it wasn't all that accessible and I was really excited," he recalled. "But to tell the truth, I was most excited to meet Mrs. Snyder. I saw her as a historical individual, part of the history of Baldwin Park." When Lozano got the chance, he told Esther how impressed he was with the facility. "Everything was well cared for," he said. "It was impeccably clean, and it obviously makes In-N-Out unique." His compliments were followed by a hesitant confession. "I told her I didn't eat meat." Lozano said that Esther smiled and laughed, and then she told him about the secret menu. "She said, 'You can request a grilled cheese.' And from that point on that is what I get to this day."

Rich needed to build more than an elegant new headquarters to execute his vision. In-N-Out Burger was a small operation. When Harry ran the chain, the corporate hierarchy consisted of just him and Esther. After he died, the chain of command expanded slightly: there was Rich, the president; his mother, Esther, the secretary-treasurer; and brother Guy, the executive vice president. Rich created a human resources department, a financial department, and for the first time, a small advertising section.

Intensely focused, Rich began assembling a new management team. "I want to grow In-N-Out," he began telling friends—he told a handful, "I want you to come and work with me."

Rich well understood that if he wanted In-N-Out to grow, he needed professional managers, and he began hiring executives who had not come up through the ranks working at the stores. During this time, he brought in a group of men with college degrees who had already gained some management experience elsewhere, men like Roger Kotch, a graduate of business administration from California State Fullerton, who started as an accounting manager (two decades later he was the chain's chief financial officer); Ken Iriart, vice president of human resources; Steve Tanner, chief financial officer, who earned his bachelor's degree in accounting from Brigham Young University; and later Carl Van Fleet, a U.C. Berkeley graduate and former Pizza Hut executive. At one point, Rich poached an executive from McDonald's, but apparently his ideas (and attitude) didn't mesh with In-N-Out's culture, and his tenure was short-lived. With few exceptions, the group that Rich put together proved to be a remarkably loyal, deep bench of talent. Nearly all stayed with the chain for decades, seeing it through a number of critical junctures.

Around the same time, Rich decided to establish a new department devoted to real estate and development. It was his desire to roll out anywhere between five and ten new stores a year. Up until then, In-N-Out Burger had used its small maintenance department to build new stores—during the day, they worked on routine maintenance, and after hours they worked on new store construction. This setup severely limited the chain's ability to build more than one or two new stores a year. If In-N-Out was to move forward, it needed to increase its ability to survey locations and erect new stores. The Snyders believed that quality control should extend to all areas, and the building of new stores was no exception.

Rich tapped longtime friend Richard Boyd, a local contractor who owned his own business and had worked on a number of In-N-Out's stores as well as the rebuild of the warehouse and new headquarters, to head up the department. Very quickly, Boyd developed a formula for construction that mirrored the system that the chain had with its food suppliers; it was based on establishing long-term relationships with contractors. As In-N-Out expanded, so did its real estate and development department. Under Boyd, the chain now had the capacity (at least in terms of construction) to build a minimum of ten new stores a year.

Under Harry, one of the cornerstones of In-N-Out's limited growth strategy was the determination to expand only as quickly as its management strength would allow. In order for Rich to be able to realize his goal of significantly expanding the chain, he needed to come up with a process to rapidly develop a strong, competent line of managers. By promoting solely from within, In-N-Out was able to preserve both its exacting standards and unique culture— but that very same system of rolling out a new store after someone had moved up to manager severely limited the chain's ability to grow substantially.

Rich was constantly on the lookout for management talent. He established incentives for store managers to find new management trainees. While eating out in a restaurant, if one of the staff caught his eye, Rich would strike up a conversation. If he liked what he saw, he'd say, "Hey, you've got a great personality. Why don't you come work for In-N-Out?"

It wasn't enough. There was another crucial element that Rich needed: the ability to increase the number of well-trained store managers as well.

CHAPTER 12

In 1984, Rich founded In-N-Out University, a large-scale management training program on the corner of Francisquito Avenue and Vineland, across from store Number One. It happened to be located on the site of Harry and Esther Snyder's former house. "If you are going to grow your organization," explained William Martin, who devised the University's initial training manual and curriculum, "you need a training program, and that's the bottom line. That was Rich's motivation for the University. He really understood that." Originally built as a full-service drive-through restaurant (if bigger than a typical In-N-Out store), the University was where management-level employees could receive instruction and learn how to run a unit in a real-world environment.* It also allowed the company to reinforce its own particular methodologies and strategies.

The concept was not entirely new. In 1961, McDonald's had founded its own Hamburger University in the basement of one of its restaurants in Elk Grove Village, Illinois. A full-time training center, Hamburger U was originally set up as an instructional program for its licensees, intended to ensure uniformity in every aspect of the McDonald's system. As McDonald's grew, its university grew as well.

---

* In 2004 the company built a bigger university, a two-story Spanish mission-style building, when it built the new store Number One on the opposite side of the university's parking lot.

By 1983, the company had moved from the Elk Grove Village basement to a new $40 million facility instructing five thousand students from 119 countries.

In-N-Out University was as practical as it was bold—at least in terms of sheer size, the Snyders' chain could not equal its early Southern California rival. The year that In-N-Out University was established, there were about thirty stores pulling in an estimated $30 million in revenue. In contrast, during the first quarter of 1984 alone, McDonald's had opened forty-one new stores and reported record system-wide sales of $2.2 billion for the quarter. By the time that Ray Kroc died in January 1984, almost eight thousand pairs of golden arches circled the globe in some thirty-one countries, generating $8 billion annually. The ledger of the two companies' numbers listed side by side might have seemed horribly mismatched, save for one figure; In-N-Out Burger was meeting the industry giant's revenues on a store-to-store basis. In something of a surprise for the notoriously tight-lipped company, Rich himself admitted as much when he told a local reporter that in terms of volume, the average In-N-Out location came "pretty close" to the volume at an average McDonald's.

According to William Martin, who had previously directed the training and organizational development at another California restaurant chain, the Snyders gave him a great deal of autonomy to come up with a program. "They didn't have an in-house, in-depth expertise for training." When he started, Martin discovered a successful company with a special culture, one that was insulated and private but seemed to vibrate with happiness. He found Rich and Esther Snyder "absolutely wonderful." Rich, he said, was "the ultimate great boss," and Esther was "gentle, kind, and lovely . . . a saint." Martin added that, "whole graciousness permeated though the company." Guy, however, was rarely around.

Initially, In-N-Out University's program consisted of a once-a-week afternoon course conducted over five weeks. Later, the training expanded to classes of about twenty students who spent 165 hours in the classroom. Martin incorporated such management staples as *Your Attitude Is Showing* by Elwood Chapman and *The One Minute Manager* into the curriculum. "It was a natural because they were so attitude-oriented," he said. The curriculum stressed how to hire, interview, and train. Perhaps not surprisingly, the program was intensely focused on areas that mirrored key aspects of Rich's personality: communication skills, methods for motivating associates, and positive attitudes. It was also tailored to help associates spot potential as well as to give them a sense of challenge and decision-making. One of the basic tenets taught at the University was called rule number one: "The customer is always right." Martin recalled that "Rule number two was, 'If in fact the customer was not right, refer to rule number one.'"

Although flipping burgers and dressing buns might be tedious, the Snyders made sure that working at In-N-Out was not. While those at the store level were expected to rigidly adhere to procedure, according to Martin the company had a real "respect for creativity and judgment." Rich gave a great deal of latitude to his employees, rewarding and encouraging those associates who showed initiative and independent thinking, particularly at the corporate level.

While Rich was eager to accrue more managers to facilitate In-N-Out's expansion, it was not easy to join their ranks. In order to attend the University, an associate usually had to have already worked full-time at a store for a year before she was eligible to start at the fourth rung as an entry-level manager (In-N-Out has a four-tier manager system in each store). During that time, the associate had to demonstrate a track record of hard work, responsibility, and potential; he also had to show that he could make good decisions, take initiative, and exhibit considerable people skills with both the customers and his co-workers. As positions opened up (and new stores opened), there would be upward movement to the next level of manager, and each level had to spend time at the University once they were promoted for additional training.

On average, it took about five years to reach store manager from entry-level management—and within the company it was viewed as a significant accomplishment. "There was no overnight promotion," explained Martin. "You had to prove yourself."

Notably, only store managers manned the grill. Unlike most fast-food chains, the company considered a grill position a highly skilled job. After all, it was the altar upon which the whole enterprise rested. It was a very intricate operation, since every single burger was made to order—a beef patty did not go down until an order ticket went up. It required a tremendous amount of coordination and speed, requiring three to six months just to learn to operate and manage.

Teaching and reinforcing consistency of quality was a real obsession with Rich, and at the University he came up with a number of schemes to maximize the opportunity to do so. The company began sending its top executives to Cal Poly Pomona College, where they took courses in a number of areas such as human resources and sexual harassment. A team of field specialists were deployed to further motivate and train in-store managers; they helped with everything from the proper way to talk to customers and associates to flipping hamburgers. Inspired by college and professional sports teams, Rich began producing a series of training films, videotaping trainees for the express purpose of critiquing their performance.

"Rich was always a leader in communication," recalled Jack Williams. "And by communication I mean really good follow-up. He developed a package that was clear in what you were expected to do, you were given training to do the job, and then there was always follow-up to see how you were doing it and if you could do it better. If you were strong in one area and not in others, they would work with the people. They gave them direction and motivation. They had an awesome selection process, and his management development and training program was never-ending."

As a result of his dyslexia, Rich had relied heavily on verbal and visual communication techniques—it was one of the reasons he launched In-N-Out's *Burger Television*. The program was a modern, company-wide motivational tool, full of colorful graphics, a vibrant soundtrack, and MTV-style quick cut edits. Similar to a network magazine program, the show broadcast In-N-Out news and events each month to the company. Often a be-suited Rich, with a *BTV* mic in hand, was shot interviewing various associates. *Burger TV* allowed the company to reaffirm its basic concepts while rallying the troops and bolstering morale.

Before long, the company launched its own newsletter called *The Bulletin.* Initially, it was a one-page, black-and-white internal dispatch; distributed to the associates, it offered details about operations and company policy. As the chain grew, *The Bulletin* expanded to an eight-page color glossy filled with breezy stories about In-N-Out goings-on (store openings and promotions) as well as the associates' personal milestones (weddings, births, graduations). Although publicity-shy, the chain wasn't averse to seeing its own name in internal headlines, and one section in *The Bulletin* was later devoted exclusively to recounting where and when In-N-Out Burger had appeared in the news.

At one point in the midst of Rich's planning, building, and growing frenzy, he decided to seek the advice of a hard-charging food industry consultant. Given his penchant for surrounding himself with mentors and his desire to constantly upgrade his own management understanding and abilities, it wasn't surprising that he sought outside counsel—but the consultant's recommendation took him by surprise. Apparently, the consultant told Rich that if he slashed salaries, In-N-Out could save a "ton of money"; the very idea infuriated Rich. This contradicted the very foundation of In-N-Out's philosophy and its success. When Rich sourly recounted the story, he said the suggestion was exactly the kind of advice one would get "from a guy who wears a suit and who thinks you don't pay a guy who cooks hamburgers that much money."

Like his father, Rich shared in the belief that running a successful fast-food business was not about cutting corners or about purchasing the right equipment. What it boiled down to was people management. Where the two differed, however, was that while Harry had hoped that his associates would work hard at In-N-Out, save their money, and then move on—perhaps even to open their own fast-food businesses (as a few of them did)—Rich had a different vision. As president, one of his chief goals was to build a much bigger footprint for In-N-Out Burger. And his philosophy was, "Why let good people move on when you can use them to help your company grow?" But he had no intention of cutting costs to inflate paper profits.

From the start, In-N-Out paid its employees more than the going rate (associates always made at least two to three dollars above minimum wage) and was an early practitioner of profit sharing. Under Rich, In-N-Out went further, establishing an expansive set of benefits under which part-time workers received free meals, paid vacations, 401(k) plans, and flexible schedules. Full-time associates also received medical, dental, vision, life, and travel insurance.

In fact, after the state of California raised its minimum wage in 1988 from $3.35 to $4.25 (its first increase in seven years), the *Orange County Register* called Rich Snyder perhaps the only restaurant executive in the state to favor a widespread pay hike. At the time, Rich had already boosted In-N-Out's starting wages to $6 an hour from $4.25. "If you lose your workers, you lose your customers," he said. "I don't know how others do it, but we just try to keep everybody happy that works for us."

Low-wage employers, particularly in the fast-food industry (which traditionally had both the highest proportion of minimum wage workers and the youngest employees), had long opposed increasing the minimum wage. Chief among their reasons was the belief that it would have a negative impact on employment. Famously, around the time of President Nixon's 1972 reelection campaign, a number of McDonald's franchisees were among a group of small businesses that lobbied Congress to prevent an increase in the minimum wage and even sought to have legislation passed that would exempt part-time students from earning the minimum wage. Critics of the move quickly dubbed it the McDonald's Bill. The fast-food chain's detractors were further angered when it was revealed that Ray Kroc had personally (and separately) donated $250,000 to Nixon's reelection campaign.

To put it in further perspective, fast forward to February 2008, when starting pay for all new In-N-Out associates (including part-time associates) reached $10 an hour. Two years earlier, the chain raised its own minimum wage for part-time workers to $9.50. At the time, the minimum wage in the state of California was $6.50 (in January 2007, the minimum wage increased to $7.50). By contrast, Wal-Mart, a company with $375 billion in sales (some ten times greater than In-N-Out's annual revenue) was paying its full-time hourly workers $10.51, only 99 cents more per hour than In-N-Out was paying its part-time hourly employees.

In-N-Out's store managers (about 80 percent of whom began at the very bottom, picking up trash, before moving up through the ranks) earned salaries equal to if not greater than most college graduates. By 1989, top store managers earned about $63,000 and were eligible for monthly bonuses tied to store performance. On average, they had been with the chain for about ten years. Accounting for In-N-Out's team of dedicated managers, Esther once proudly stated, "We're blessed with good employees, who run the restaurants as if they were their own stores." Certainly, their high salaries went a long way toward explaining their longevity. Some twenty years later, store managers were pulling in at least $100,000 annually.

It was Rich's belief that his job was the bottom point of an inverted triangle. He was there to support everyone else in the company. When talking to store managers, he was always careful to refer the shops as "your stores" and never asked them "What store do you manage?" He wanted them to have pride of ownership. Regardless of anyone's position or length of time with the company, Rich treated everyone equally and as if they were all special. "He really lived the Christian belief of treat your neighbor as yourself," exclaimed Rich's good friend Heath Habbeshaw. An ordained minister, Habbeshaw began working at the La Puente In-N-Out as a teenager with Rich in 1968 and remained with the company on and off until the mid-1990s. "He poured that into his business philosophy and everybody loved him. It made us all work even harder."

As a result, In-N-Out could also boast one of the lowest turnover rates in a high-churn industry. According to various analyses, in the fast-food world, little more than half of the workforce stays behind the counter for one year or more, with roughly 75 percent of employees staying on beyond six months. After that, the numbers decrease substantially: 53 percent remain one year, 25 percent stay two years, and only 12 percent remain three years or longer. In the case of In-N-Out Burger, its managers maintained an average tenure of fourteen years, while its part-time associates remained, on average, two years.

The result was a corporate culture operating in stark contrast to the competition's systems of burger flippers and vat fryers, floor moppers and cashiers who put on their paper hats and grease-stained aprons in what society calls McJobs and economists refer to as the requisite churn of capitalism. It was a place where people genuinely enjoyed getting up in the morning and going to work. Rich explained it this way: "We try and maintain the highest quality level possible, and to do that you need good training and good people. That's why we pay the highest wages in the industry." He added, "It means we tend to keep our employees longer than at other places, and the reduced turnover helps us maintain consistency in our products." Notably, the philosophy did not trade on or lead to either higher prices or lower-quality food.

For its part, In-N-Out was selective in its employment process. When hiring, the chain probed a potential candidate for her ability to not only meet but exceed customers' expectations while working as a team. Interviewers asked detailed questions, zeroing in on the candidate's view on interacting with others, looking for signs of flexibility and the ability to deal with people holding a spectrum of opinions. Just as Harry frequently told his son, Rich looked at his associates and said, "I believe in you. You are the best."

The associates were considered the chain's front lines. For starters, all were required to keep up a clean-cut appearance. All hires were expected to maintain a friendly attitude toward customers (or, rather, "guests"), smiling and looking them straight in the eye. "Times are tough," he told them. "People are going through a lot. The only smile or friendly service they get might be the one you give them." Rich felt so strongly about it that he launched a "Smile of the Month" feature on *Burger TV* to recognize the associate who demonstrated the best smile.

New hires started at the bottom, picking up trash, wiping counters, and putting orders in trays for customers. After proving that they had mastered their current assignment, associates moved up the ladder to filling beverages, dressing burgers, and frying potatoes. The counter associates were instructed to always repeat orders, ensuring that each one was absolutely correct. "They didn't just hire anybody," recalled Russ Nielson, who worked part-time at the Hesperia store for one year when he was sixteen years old in 1990. "They wanted to make sure that you had moderate intelligence and were above average. I remember there was constantly an influx of applications all of the time. Everybody wanted to work there. It was good money, and you could eat for free, too."

Associates were never hung out to dry. They were given specific on-the-job training during slow periods and a considerable amount of feedback on their performance. The point was to make sure that each associate understood his job and how he could do better, and associates were given more of a customer load and more responsibility incrementally and according to their abilities. The elapsed time between starting in cleanup and working the french fry vats could be as long as a year and a half.

Although the work could be monotonous and dreary, four-hour shifts of cleaning ketchup spills or doing nothing but filling soft drink cups with soda, associates were made to feel that they were part of an important enterprise and all were given the opportunity to advance. At In-N-Out, they had a future. There were numerous part-timers who joined In-N-Out for a summer or as an after school job and stayed on, becoming store managers, moving further up the corporate management ranks, and making lucrative careers at the burger chain. Like his father, Harry, Rich expected much from his associates, but in turn he treated them extremely well and continued to offer them opportunities to continue their education and expand their skills and talents.

In order to maintain the chain's strict quality standards even as it grew, Rich implemented a small army of "secret shoppers." These undercover customers went from store to store on a monthly basis, making sure that associates were properly dressed and clean, orders were correct, food was presented properly, and even that the right amount of change was given. Sometimes they'd order complex meals to see whether they could trip up an associate. But associates never had any idea until afterward. If they performed badly, they were informed and given opportunities to do better, unless their performance was egregious. An exceptionally good performance was usually rewarded with a small cash bonus. In essence, the secret shopper system was really just a larger-scale variation on Esther and Harry's frequent store visits. As a former associate put it, "They were mostly looking for perfection."

Rich earned a reputation as a passionate leader who led by personality. A consummate professional, he always had a smile on his face and seemed to effortlessly convey a deep emotional commitment to In-N-Out that extended to all of his associates. Like his parents, he visited the stores often and chatted up the associates whom he still made a point of knowing by name. His connection to the company was matched by his deep feeling for everyone who worked there. As one of his friends later recalled, "He never acted like the boss. I remember once I made the mistake of calling the workers 'employees' instead of 'associates,' and he corrected me instantly." On several occasions, Rich asked colleagues to look up the words "associate" and "employee" in the dictionary. Then he would say, "I'd rather have an associate work for me than an employee."

Rich set the tone for the company. Each year, he held a company-wide picnic for all associates and their families. The picnic was spread out over two days so that all In-N-Out shifts could be accommodated, and the company chartered buses that ferried the attendees to the picnic grounds. Once there they found an unlimited supply of food (all gratis) and games and activities such as hot air balloon rides. And there were always raffle contests in which TVs, trips, video games, and the latest technological gadgets were available as prizes. There were trips to Knott's Berry Farm as well as Halloween parties for associates and staffers and elaborate Christmas parties where, on occasion, Rich would truck in real snow, to the delight of the associates and their children.

At the start of each year, Rich threw a black-tie, gala dinner for associates often held in one of the ballrooms at the Disneyland Hotel in Anaheim. The night before the gala, Rich held a kick-off dinner for all of the managers and any associate who wanted to attend. During these occasions, he usually invited a slew of marquee names from a variety of areas—particularly professional sports—to attend. Through the years, Dodgers manager Tommy Lasorda, Lakers head coach Pat Riley, and the basketball team's executive Jerry West gave motivational speeches to In-N-Out's top managers. "We want them to share some of their insights," Ken Iriart, the chain's vice president of human resources, once explained to the *Los Angeles Business Journal*. "Some of these people went into good detail about what went into building a winning team, and some of these same things apply to business."

The gala was a combination fete, cheerleading session, award ceremony, and black-tie affair, all rolled into one. At the dinner, the Snyders handed out the Harry Snyder Award—In-N-Out's version of employee of the year—to its most outstanding associate. It was the time when new managers were named. Another hallmark of the annual gala was the announcement of the location of the upcoming year's 100 Percent Club trips. Esther Snyder loved to travel, and a few years after Rich had begun running In-N-Out, they came up with a special program to reward managers when they reached their goals. Those named to the club (along with their spouses) were awarded first class trips to such places as Hawaii, the Caribbean, Australia, and Europe. During the course of the trips, Rich often invited big-name speakers to continue to inspire the managers.

In his mind, In-N-Out managers were just as important as the executives at any Fortune 500 company. That's why Rich created the annual gala; it's why he took his executives to numerous cultural events. At a Christmas time performance of the *Nutcracker* ballet, Rich required his managers to wear tuxedos. Rich thought they stood shoulder to shoulder with any blue chip manager, and he wanted them to feel similarly.

CHAPTER 13

Soon In-N-Out Burger began attracting serious attention. Over the years, the media-shy Snyders had given very few interviews, and those few were usually with the *San Gabriel Valley Tribune*, which covered the chain largely as a popular local story. There were of course various mentions and articles in trade publications, but the Snyder family participated only infrequently. Before long, however, the In-N-Out phenomenon had become hard to ignore. In 1989, *Forbes* magazine featured In-N-Out in a glowing article entitled "Where Bob Hope Buys His Hamburgers," declaring that the "anti-fad has become a fad." For the publication that bears the motto "Capitalist Tool," Rich Snyder not only agreed to be interviewed, but the smiling In-N-Out president was photographed in shirtsleeves and a tie, holding a cardboard box stuffed with a Double-Double and french fries. Although the family did little to encourage further publicity, soon after the *Forbes* article, a passel of stories about In-N-Out Burger began to appear.

The media attention certainly helped to generate a new level of interest in the small regional chain, but nothing gave it more publicity (or sales) then its own longtime customers. Staying simple and remaining focused on its core values had allowed In-N-Out to stay true to its loyal fan base. And it was precisely those customers who often did the heavy lifting, frequently boasting about their zealous affection for the chain to everybody else. Regulars (who almost always ordered off of the secret menu) engaged in an ongoing contest, trying to outdo each other on how many hamburgers they could eat at any one time. Colorado native Duke Sherman proudly recounted the episode that occurred when he was studying for final exams at Occidental College in Eagle Rock, a northeastern Los Angeles neighborhood. "Four of my dorm mates—two football players, a violinist, and a rugby player—decided we would make In-N-Out history," was how his story began. The plan was to head to the nearest In-N-Out in Pasadena, about five miles from campus, and each attempt to consume twenty-five Double-Doubles in one hour. "We made quite a scene that night as we ordered and ate with great fervor," he recalled years later, after moving to New York and launching his own consulting firm. "I think we would have been more successful had the staff not provided free fries to help get us in the spirit. But the bacchanal did not last long, and we set no records. We left In-N-Out men that night, men who were complete, men who were nauseous."

Such ignominious tales of ingenuity and one-upmanship reached an apogee many years later on Halloween night in 2004, when a group of eight San Francisco friends famously ordered and finished a 100 x 100 (one hundred beef patties and one hundred slices of cheese) at a Las Vegas In-N-Out. The tab for the monster cheeseburger: $97.66.

The swapping of stories about the lengths to which one would go in order to eat at an In-N-Out further fueled the chain's mystique. Richard Clark, the owner of Clark Copper Head Gaskets in Minneapolis, boasted of paying over one hundred dollars in freight to fly forty Double-Doubles from the Ontario In-N-Out for his son's twenty-first birthday. A former dragster, Clark ate his first In-N-Out burger in 1971 while he was on the circuit and still gets misty-eyed when he talks about the chain. "They're just the greatest hamburgers ever made," he trilled. "Whenever I fly into California, I get the rental car and then go to the In-N-Out. Even before I check into my hotel room, I'll eat a couple of Double-Doubles at one sitting. If I'm there five days, I eat one every day." Clark sighed, adding: "My son-in-law is out West, and when he goes to In-N-Out he'll call me and say: 'Hey I'm eating a Double-Double,' and hang up just to piss me off."

As its competitors looked on with envy, the chain's regulars assumed the responsibility of bringing in a constant stream of new devotees, an act generally referred to as "the conversion." The deed had the feel of bestowing membership into a club that seemed at once exclusive and egalitarian.

The prototypical conversion story goes something like this: "I was one of the converts," proclaimed Angela Courtin, a marketing executive in Los Angeles. "I moved here from Texas. My brother was already here, and he said, 'You have to try In-N-Out.' After that, I started going once a week. I love the simplicity of the menu, and it's fresh. A friend of a friend let me in on the secret menu. After that, I felt I must pass this on and tell other newbies—and I've done that for numerous people. Actually, I feel it's akin to my civic duty. I've always said that In-In-Out is the perfect blend of communism and democracy. There are limited choices, but it is an entrepreneurial family business. It is the great class equalizer. Look inside! You get everybody here: middle-class skateboarders and Beverly Hills ladies, ethnic families and day laborers, all eating in the same restaurant, at the same price point, and with the same three options."

Master chef and American food pioneer Julia Child, a California native, counted herself a longtime fan. Nancy Verde Barr, Child's friend and assistant, wrote fondly in her own memoir, *Backstage with Julia: My Years with Julia Child*, about the day when the two "gobbled down Double-Double burgers at the In-N-Out drive-through" near Child's Santa Barbara home. Indeed, the woman whose kitchen was eventually displayed in the Smithsonian Institution was said to carry a list of store locations in her pocketbook.

A Hollywood favorite from its first days, In-N-Out had in the intervening years become as much a part of celebrity culture as autographs or the red carpet. Without losing any of its homespun appeal, the burger chain's popularity among the in-crowd was entirely the result of a wide swath of celebrities who regularly talked it up. "When I first joined the band, we must have eaten there at least three days a week," recalled rocker Sammy Hagar, who signed up as the front man for Van Halen in 1985. "We were in the studio recording *5150*, and we'd send someone to go get food, and we'd talk about sushi or pizza and always end up with In-N-Out." Gordon Ramsay, the British celebrity chef with twelve Michelin stars, global fame, and profanity-laced rants, once admitted to sitting down for a Double-Double and then "minutes later I drove back 'round and got the same thing again to take away." PGA golf champ Phil Mickelson mentioned the chain so often that whenever he fell into a losing streak, sportswriters began suggesting that he cut back on the Double-Doubles.

Before long, tourists got wind of In-N-Out Burger and began making their own pilgrimages to what was considered the quintessential Southern California attraction. Fans passed the secret menu on to one another and described the sublime pleasures of tucking into an Animal Style cheeseburger. Vegetarians talked up the chain's off-menu Grilled Cheese. Expatriate Californians pined for their favorite burger, and In-N-Out T-shirts were the epitome of cool. Analysts spoke of In-N-Out's "uncopyable advantage," while everybody else talked about its unparalleled cult following. According to William Martin, the Snyders and the rest of the chain's highest echelon were definitely conscious of the mystique that had developed around In-N-Out. "They were all aware of it, and they loved it," he said. "But they had no explanation for it." That didn't mean, however, that they didn't know how use it.

In another radical departure from Harry, Rich began implementing a more aggressive approach to marketing the chain. In 1977, only a year after he took over, Rich launched In-N-Out's first television commercial. The brief animated spot featured a cowboy astride a horse riding up to an In-N-Out and ordering a Double-Double. Under Rich, the chain launched a radio jingle with the catchy refrain: "In-N-Out, In-N-Out, that's what a hamburger is all about," and it fast became something of a Southern California anthem. Despite the high-voltage marketing, Rich was careful to ensure that its message—like the company itself—remained focused on the product. It did not, as had become commonplace among its competitors, revolve around some kind of manufactured experience. Before long, a pair of new slogans, "Quality You Can Taste" and "Cleanliness You Can See," appeared. Consumers felt a connection with In-N-Out's simple, almost quaint brand, and it was clear that Rich knew the value in preserving that.

For some time, In-N-Out's signature logo had served as an important advertising feature. Echoing the days when roadside diners attracted travelers with their kitschy neon architecture, In-N-Outs strategically placed along freeway off-ramps served much the same purpose. The yellow boomerang arrow was a beacon for weary and hungry travelers. As it turned out, In-N-Out's real estate was one of its best marketing strategies. Continuing with that theme and playing up its roots as a drive-through, the chain began printing up maps pinpointing stores and their addresses for its customers. In time, In-N-Out produced a compact, pocket-sized location booklet, later setting up a toll-free number. In addition to asking questions about everything from the amount of fat in a cheeseburger (twenty-seven grams) to what kind of oil the french fries are fried in (100 percent vegetable oil), drivers on the road could call in and tell an operator where they were and be directed to the nearest In-N-Out. (That feature was updated in later years when In-N-Out installed a map finder on its website that plotted all the In-N-Out stores in any given location.)

Reaching back to In-N-Out's early days when roadside diners were exploding and transforming the landscape, the chain deployed a deceptively simple billboard strategy that traded heavily on the chain's postwar image. Intentionally placing the large signs on streets leading to an In-N-Out, they often said little more than "In-N-Out Burger 2.5 Miles Ahead." Usually the billboards were accompanied by a three-dimensional, oversized replica of a Double-Double. When an ad agency came up with the idea to tempt drivers further by having artificial steam rise from the burger, Rich eagerly signed on.

One of In-N-Out's most successful marketing strategies came in the form of bumper stickers. In Southern California, starting in the early 1980s, placing an In-N-Out sticker on the back of one's car signified membership in a peculiar sort of club; all along the freeways, horns were honked, thumbs were raised, and heads were tipped in recognition.

At one point, it became common practice among young men across the Southland to excise the "B" and "R" from the word B-U-R-G-E-R, modifying the sticker to read "IN-N-OUT URGE"; the clean-cut company was not amused. As a result, the chain discontinued the original sticker and printed up a new one. This time a well-placed image of a Double-Double was placed on the spot where the word "burger" once stood.

In 1984, Rich used the widespread popularity of the chain's bumper stickers to launch one of In-N-Out's largest and most successful promotional initiatives: a chain-wide sweepstakes contest. During several month-long periods, In-N-Out gave away its sought-after bumper stickers, and spotters were instructed to jot down the license plate numbers of cars bearing the chain's stickers. The numbers were then entered into a series of drawings. Prizes included trips to Hawaii, microwave ovens, video recorders, and In-N-Out T-shirts. Advertised on billboards all over the greater Los Angeles area, customers helped the campaign with a grassroots effort of their own, passing the word along. In a town built on dreams, cars, and sunshine, the promotion became one of those curious Los Angeles cultural events wherein eating a burger might win you a moment of fame and a trip to the beaches of Waikiki. (Decades later, buttons and other ephemera from the original sweepstakes were sold on eBay)

The promotion combined two of Los Angeles's cultural icons (hamburgers and cars), played up In-N-Out's core values, and was relatively inexpensive to put on. It also packed an incredible public relations wallop. As the trade publication *Nation's Restaurant News* marveled at the time, "The burger Goliaths doing business in Southern California must surely envy the kind of hometown customer enthusiasm which can turn little David in the form of In-N-Out into a self-advertising car cult."

On its home turf, simply through word of mouth and bumper stickers, In-N-Out had become as well-known as Disneyland or McDonald's. Reported estimates on In-N-Out's advertising budget ranged from $1 million annually, increasing later to between $5 and $10 million—a drop in the bucket compared to the budgets of its fast-food rivals, who routinely poured hundreds of millions of dollars into their own yearly campaigns. With a continual demand for In-N-Out's burgers, the chain had little need for the kind of sophisticated multiplatform campaigns that its rivals regularly employed. In-N-Out relied on its radio jingle and the infrequent television commercial (usually broadcast on cable). Astonishingly, its billboards, bumper stickers, T-shirts, and its own rabid fans seemed to broadcast the message to great effect. Without corporate solicitation, a roster of celebrity names regularly endorsed the chain.

Straightforward and uncomplicated, the chain's marketing efforts really added up to an awareness campaign that casually reminded people about In-N-Out. As Robert LePlae, the president of TBWA/Chiat/Day, the advertising giant behind memorable campaigns for Levi's, Taco Bell, and Apple Computer, acknowledged, "Their marketing is really brilliant. The best marketing is word of mouth, and they have that. You can't get that through traditional media." Moreover, LePlae was full of praise for the simple fact that In-N-Out hadn't sold its soul. "They don't abuse the privilege that they have built up with their customers." While heading up the agency's Los Angeles office between 2001 and 2005, the self-confessed In-N-Out fan—who fondly recalled ordering cookout trailers for big agency parties and taking out-of-town clients to In-N-Out—says he learned to master the art of the "one hand on the steering wheel and the other one on the Double-Double." "They haven't commercialized the secret menu," he enthused. "There is a powerful trust between the company and the customers that is deeply ingrained. I'm not sure if it was intentional, but it is not the kind of thing a big, massive company could do. They would merchandise every little thing."

While In-N-Out's profile rose considerably, the Snyders remained guarded, at least publicly. Rich clearly saw the importance of press coverage, but stayed somewhat wary of the media. The chain adamantly

refused to discuss its operating strategies or sales figures, hewing literally to the meaning of the phrase "private, family-owned company."
Unlike the owners of many other successful businesses, the Snyders
did not relish the idea of having their portrait on the cover of *Fortune*
magazine or their benchmarks featured in the *Wall Street Journal*. Perhaps naively, the chain did not want to come across as if it were actually seeking publicity.

Indeed, in an increasingly solipsistic industry, In-N-Out remained the exception and not the rule. Take the example of Dave
Thomas, the founder of Wendy's Old-Fashioned Hamburgers (and a
former Kentucky Fried Chicken franchisee). Starting at the tail end
of the 1980s, Thomas became his chain's TV pitchman. After appearing in over eight hundred commercials between 1989 and 2002,
Dave Thomas became a household name. At one point, Wendy's conducted a survey, the results of which claimed that 90 percent of all
Americans knew who Thomas was—more than could identify the
prime minister of England. But back in California, there were few
who couldn't recognize the tell-tale symbols of In-N-Out. In fact,
although "In-N-Out" is not printed on the chain's beverage cups, one
would be hard-pressed to find a Californian who couldn't recognize
the brand from the red silhouetted palm tree design.

Rich, who had earlier posed for *Forbes* with a big smile on his face
and a Double-Double in his hand, later was said to have had misgivings about the display. Specifically, in a world in which McDonald's
proudly announced to the world that it had sold "billions of burgers,"
the magazine reported that In-N-Out sold fifty-two thousand burgers
per month. It was a slip Rich quickly came to regret. A year earlier, in
1988, he had told the *San Gabriel Valley Tribune*, that he hoped to sell
27 million burgers in 1989. In retrospect, Rich thought it smacked
of braggadocio and, perhaps worse, it made In-N-Out Burger sound
like every other fast-food place. In the future, there would be scant
public mention of how many burgers In-N-Out had sold.

As the years wore on, the family retreated further from the spotlight. Increasingly, they pushed Carl Van Fleet, the vice president of
operations, into the role of In-N-Out's spokesperson. It was a posi-

tion that mostly consisted of offering firm but polite "no comments" and other genial but equally opaque statements. On occasion, Rich broke his press silence—however, this happened infrequently. Esther Snyder was rarely (if ever) heard from publicly when it came to discussing the chain, and Guy Snyder even less.

In describing the chain's rather cool attitude toward publicity, Carl Van Fleet once explained, "We aren't striving to become a household name."

Although it was most likely unintentional, this enigmatic quality added a definite layer of intrigue to the chain. It was a lesson the Snyders took to heart. As In-N-Out's onetime chief financial officer Steve Tanner once said, "If you have to tell somebody you're something, you're probably not."

CHAPTER 14

As In-N-Out Burger continued to grow and prosper, Rich found himself turning closer toward religion. He had come to the conclusion that God was behind the success of his business and religious devotion was the way to a better world. For him, business, religion, and politics overlapped. Already a believer, personal events had increasingly propelled him further toward religion. Harry's illness was a seminal event. Earlier, while traveling by plane with his father from Mexico where Harry was receiving treatment for his lung cancer, Rich prayed with him to receive Jesus. Then sometime around 1983, Rich was born again as a Christian. "He gave his life to the Lord," explained his cousin Bob Meserve, "and everything turned around from that point forward."

It was a slightly different spiritual orientation from the one that existed in the Snyder home while Rich and Guy were growing up. Although Harry held to traditional values, he didn't seem to feel much of a need for organized religion. As Harry's cancer progressed, however, he too had become more religious. According to his nephew, "On his death bed, he received the Lord."

Raised as a Free Methodist, Esther always maintained a deep and abiding faith. She believed that religion was a private and personal affair. Said to almost always carry a Bible with her, Esther didn't care what religion others espoused and she didn't always attend church, but she did believe in God. While interviewing his mother for a home movie about her life, at one point Rich asked Esther what her favorite subject in school was, to which she replied "science." Rich, who was off-camera, seemed startled by her answer. Retorting with some incredulity, he remarked, "You believe in God and you still enjoyed science?" Looking straight into the camera, her eyes widening, Esther barely hesitated. "Sure," she replied. "Even Darwin believed."

For his part, Guy Snyder considered himself to be a Christian at heart, although spiritual salvation just wasn't central to his life—but when he traveled, he liked to visit different churches. The brothers' differing levels of religious commitment and faith was just one more line of separation that pushed them ever further apart.

It was during this period that Rich became actively involved in the Calvary Chapel of Costa Mesa in Orange County. One of the first mega-churches, Calvary had grown from humble beginnings. It began in 1965 as a congregation of about twenty-five parishioners who met in a mobile home after its founder, Pastor Chuck Smith, broke away from the International Church of the Foursquare Gospel in Santa Ana.

Smith was a leading figure in the grassroots "Jesus Movement." Known to some as the "Jesus Freaks," members belonged to a religious revival born out of the hippie counterculture of the late 1960s. The movement got its start in a storefront in San Francisco's Haight-Ashbury district called the "Living Room" and spread quickly. Down in Orange County, Smith's early outreach to hippies spurred thousands to flock to his church. As it turned out, the Jesus Movement became a spawning ground for what was to become another Christian phenomenon: the evangelical mega-church.

By the time that Rich became involved, the Calvary Chapel had developed into a global ministry claiming thousands of conversions. In time, the Calvary Chapel would boast over one thousand congregations as well as a television and audio ministry.

As his involvement in the church deepened, Rich's faith moved to the center stage of his life, permeating every aspect of it. "What is important to know about Rich," exclaimed Jack Williams, "is that he loved the Lord. The entire foundation of In-N-Out was built on Christian values. He didn't push Christ down anybody's throat, but he lived it and talked it twenty-four/seven."

It was a heady time for Rich Snyder. For him, the 1980s were a period of extremes. Rich had invested considerable energy in building up the family business. His personal life, however, did not seem to match his business success. His weight continued to fluctuate. When the stress got to him, he harbored a secret desire to open a hardware shop in Phoenix. Approaching his thirties, Rich remained single. One of his greatest desires was to marry and start a family of his own. While a number of women expressed interest, Rich found that most of it stemmed from the fact that he was president of In-N-Out Burger; the experience left him feeling empty. Although those close to Rich described him as chronically upbeat, he had also quietly undergone what he considered to be a set of personal trials and tribulations—and those he kept close to the vest.

Privately, Rich found solace in helping others. He remained in the lives of Kim Stites and her daughter, Meredith, who was just nine days old when her father, Wilbur, died. She called him Uncle Richie. Every Sunday, the families had dinner together, and each December Rich took Meredith on a special adventure to pick out a Christmas tree. Frequently (and often unannounced), Rich could be found visiting homeless shelters where he personally passed out blankets and other supplies. Without any kind of public announcement (and without the knowledge of his family or even the top circle at In-N-Out), Rich donated his own money to help build a chapel at the Union Rescue Mission. Located in downtown Los Angeles's Skid Row, the Mission had aided poor and homeless families since its founding in 1891 by Lyman Stewart, the president and founder of the Union Oil Company.

Even though he had none of his own, Rich had become particularly concerned with helping children. For him, it was almost a crusade—it came from his belief that improving the lives of children was the cornerstone to creating a better society. And Rich looked to fund outlets that shared his desire to help, supporting several groups that aided victims of child abuse, organizing several fund-raisers. He was said to have donated thousands of dollars to the Make-A-Wish Foundation, the organization that granted wishes to terminally ill children.

Esther Snyder shared her son's feelings. For years, she had been actively involved in supporting the Boys and Girls Clubs of America—an organization that provided children across the country with a host of educational and recreational programs—giving the group tens of thousands of dollars. She also sat on the board of the club's Baldwin Park branch. Bob Benbow, who sat on the board with Esther for a number of years starting in the 1970s, recalled how she believed sincerely that it was important to give back. "She felt blessed," said Benbow, an avuncular Texas transplant who arrived in Baldwin Park in 1961 to teach high school. "She had so much empathy. She was here during those days when there were so many young people living in the trailer camps. They were mostly migrants. She wanted to help." As always, Esther was quick to donate money when asked or to supply In-N-Out burgers for any number of charitable events. "She helped in any way she could. One time she gave us $50,000 in one whack."

With the continued success of In-N-Out and their own personal convictions, Esther and Rich decided to formalize a vehicle for their own continued giving. In 1984, they established the Child Abuse Fund specifically to raise awareness of and channel funds to abused and disadvantaged children in the cities and counties where In-N-Out Burger did business. The foundation launched with a storewide, month-long charity drive that became an annual event. Each April, canisters were placed in all of the chain's stores so that customers could contribute small bills and coins, with In-N-Out matching the funds up to $100,000 (later the company increased the amount of matching funds to $200,000).

Two years after the launch, on June 6, 1986, a photograph in the *San Gabriel Valley Tribune* showed a beaming Rich (in his customary suit and tie) along with three In-N-Out Burger executives behind a table piled with cans, stuffed with $20,329.81. "Burger Promotion Aids Abused Children," reads the headline.

Within three years, the chain's charity had grown well beyond its can collection. The Snyders added an annual golf tournament to benefit children as well. It was a huge event, usually held at a prestigious California golf club. Store managers and their associates played in teams against each other in what quickly became a widely anticipated activity. There were large sponsorships, a dinner and auction at the end of the day, and huge prizes. Over the years, the fund distributed millions of dollars.

In addition to In-N-Out's reputation for juicy burgers, the chain was earning the respect and admiration of the numerous civic and social groups that benefited from its largesse. They became good corporate citizens in each community where a new store opened. They regularly sent associates to pick up the In-N-Out trash that littered neighborhood streets and sidewalks. They sent cookout trailers to feed lines of firemen battling fires, offering assistance during times of need in myriad ways. And In-N-Out contributed greatly to a number of police and California Highway Patrol charities, especially those that helped the families of officers killed in the line of duty. Law enforcement held both In-N-Out and its burgers in high regard. Esther, who like Guy had a taste for speed, loved cruising down the highway. She said that the police viewed her with grandmotherly affection and kept an eye on her. Esther giggled to family and friends that despite her lead foot, she almost never received a speeding ticket.

Not long after Rich announced his deep commitment to Christianity, he had Bible verses printed on In-N-Out's packaging. Discreetly tucked inside the rim on the bottom of the soda pop cups, it said simply: "John 3:16" ("For God so loved the world that he gave his only begotten Son"). Shortly thereafter, other verses cropped up unannounced on milk shake cups "Proverbs 3:5" ("Trust in the Lord with all Thine heart"); hamburger and cheeseburger paper wrappers "Revelation 3:20" ("Behold, I stand at the door and knock"); and on Double-Double wrappers "Nahum 1:7" ("The Lord is good, a strong hold in the day of trouble").

The unusual marketing move was received with a combination of criticism, applause, and outright disbelief. Mixing business and religion also fueled a host of urban myths about the chain, the most common (and enduring) being that a Christian sect owned In-N-Out.* Despite the rampant speculation, the truth was straightforward; Rich decided to print the verses in an effort to share his faith through mass culture. "Hamburgers are so popular," was how Chuck Smith Sr., Calvary Chapel's founding minister, explained Rich's reasoning. "He thought it was a great way to awaken people to the fact that the Bible is relevant and has the answers to today's problems."

Around 1991, a few years after he first printed the Bible verses, Rich decided to broadcast a message of salvation on Los Angeles–area radio stations during the Christmas season. The spot opened to the familiar In-N-Out jingle, but with an orchestral rendering. A voiceover asked listeners to consider letting Jesus Christ into their lives. Although some of In-N-Out's marketing team expressed concern the move might offend the chain's non-Christian customers, Rich went forward. It wasn't his intention to upset anybody, he told them, but he had a bigger picture in mind.

Unsurprisingly, within the largely secular radio culture of Southern California, the commercial provided a short-lived holiday controversy. Some stations rejected the commercial outright, while others decided to broadcast a generic version of the spot (also provided by In-N-Out). A few stations chose to air the religious version only on Christmas day. "It gets the Christian community pretty excited seeing In-N-Out being pretty bold like that," Roger Marsh, the general manager of KYMS-FM (a now defunct Christian station in Orange County), exclaimed at the time. His opinion was not widely shared. The ad spurred scores of calls to the burger chain. When one radio listener called to complain, telling Karen Thorton, an In-N-Out spokesperson at the time, that "not every-

---

* In-N-Out's fan-conspiracists like to point to the crossed palms at each store as evidence of this. In fact, they are a nod to Harry Snyder's favorite movie, *It's a Mad Mad Mad Mad World*, in which hidden treasure is buried beneath four palms that form a "W."

body that listens to you is a Christian," she replied, "Well, that's too bad."

Rich seemed impervious to the controversy he had created. "It would be a real drag to die and be up in front of God and have to say I refused to run this type of commercial," he told the *Orange County Register*. "My love of Jesus is greater than my fear of what people will say."

His faith was the prism through which Rich viewed his stewardship of In-N-Out. "This is God's company," he said frequently, "not mine."

For Rich, everything rested on values. When it came to politics, religion, and business, he saw little distinction. And it was during this period that he became actively involved in conservative politics. Rich felt that America's spiritual foundation had eroded, resulting in a host of social woes. The Reagan administration, with its emphasis on conservative social values, limited government, personal liberty, and America's destined role as the leader of the free world battling the "Evil Soviet Empire" made perfect sense to Rich. "He thought that Reagan was a great president," John Peschong, a onetime executive director of the California Republican Party and political consultant, recalled. "He thought he was somebody that was nurturing freedom, and upheld the values that this country was founded on."

Rich's passion for America went beyond simple patriotism; it was visceral. "He had a real love for this country," recalled Peschong, who got to know Rich through a shared involvement in state politics. "He had a real sense of pride. It seemed to be part of his whole entrepreneurial spirit. He was successful because the country was successful and free." Conceivably, then, it was no coincidence that during President Reagan's second term in office, Rich began printing the motto "The Best Enterprise Is a Free Enterprise 'God Bless America'" on the company's official stationery.

Enthusiastic and filled with a deep love for Jesus, Rich was an uncommon mogul. He was dedicated to God, his family, and In-N-Out Burger. Rich felt a heartfelt responsibility to make the world a better place, and he carried that burden on his broad shoulders. It wasn't power or gaining an edge for his company that motivated his involvement in politics, but a desire to make a significant change and his core belief that Republican Party values were the way to do that. San Clemente lawyer Ken Khachigian, a longtime party activist who had been a deputy special assistant under President Nixon and later a speechwriter for President Reagan, recalled that "Rich felt that the values espoused by the Democrats and the liberals were inimical to his own personal values and standards."

For Rich, America was the greatest country on earth in which to live and work, and he felt obligated to do everything he could to support it. In 1980, he backed Republican David Dreier of the San Gabriel Valley during his inaugural congressional campaign with a $1,000 contribution. A staunch conservative, Dreier established his Republican bona fides as a supporter of tax cuts and President Reagan's anti-Communist foreign policy. Soon Rich began contributing tens of thousands of dollars to Republican candidates and causes. In time, he became a member of the exclusive "Team 100." The group of Republican donors—also known as "T-100"—all agreed to make an initial $100,000 contribution to the Republican Party in order to join. (A minimum donation of $25,000 for three years was needed to retain membership.)

Around 1985, Rich became acquainted with Bruce Herschensohn, a Republican commentator in Los Angeles. Herschensohn had been a speechwriter in the Nixon White House, and the two men came together over their unstinting support for the Reagan administration. "He heard one of my speeches and I loved In-N-Out Burgers and we became friends," recalled Herschensohn. "When it came to politics, we agreed on everything."

In 1986, Rich encouraged Herschensohn to run for the U.S. Senate against Democratic senator Alan Cranston of California. Cranston, who had been an unsuccessful presidential candidate in 1984, was seeking his fourth term in Congress. Rich contributed $10,000 to Herschensohn's exploratory committee and served as its chairman. According to Herschensohn, Rich was extremely influential in helping him network with the powerfully connected, especially at big functions. "I didn't like to interrupt people, and he'd walk with me and say 'Hello, I want to introduce you to Bruce,'" he recalled.

Herschensohn called Rich "a vital force for my running for the Senate," especially when it came to the act of fund-raising, an activity with which Herschensohn was particularly uncomfortable. After a few awkward incidents, Herschensohn finally told Rich that he had a problem asking people for money. It stemmed from the time his mother had chided him for asking his grandmother to buy him a chocolate Santa Claus at Christmas. As he recounted, "My mom told me that she didn't have that much money and it hit me hard. When I told Rich, he stopped asking me to ask people for money during the campaign." And on each of Herschensohn's subsequent birthdays, Rich sent him a chocolate Santa Claus. "Where he got them, I have no idea, since it was September," he laughed. "I still keep one in my freezer."

Esther was often found alongside her son Rich; she accompanied him to numerous political events and various In-N-Out–related functions. They frequently traveled together. She appeared to hold Rich in the same esteem as she had held her husband, Harry. He became her rock after Harry's death, and the two shared an incredibly warm and loving relationship. Rich was particularly devoted to and protective of his mother, calling her every day. Concerned that his frugal mother continued to drive an aging car, Rich surprised her by buying a new Lexus that she discovered when she opened her garage door. Esther always called him "Richie," and he liked to gently kid her. When he was a child, she told him to always smile, and in his adulthood, he taught her how to shake hands with people. In 1993, Rich delighted Esther by having a plaque commemorating her wartime service displayed at the U.S. Navy Memorial in Washington, D.C.

About ten years after Harry died, Esther moved from the San Dimas home that she had shared with her husband to the suburb of Glendora six miles away. Although she loved her house, Esther moved out, largely at Rich's insistence, who preferred that she live in a one-story house rather than aggravate her bad knees climbing stairs.

In a sentimental gesture, he had the door from her childhood home installed in her new house. An affluent residential suburb of shady oak trees at the foothills of the San Gabriel Mountains, Glendora was located on historic Route 66. The Snyders' San Dimas house and the rambling property where Harry once conducted business meetings and where Esther had for a time run the accounting operations for all of their enterprises were later sold to Christ's Church of the Valley. As part of the deal, Rich insisted that the Snyder house be left intact on the property. And there, tucked behind the boxy modern church complex, the family's two-story stucco home remains to this day.

Officially, Esther was In-N-Out's secretary-treasurer. Unofficially, she was the chain's undisputed emotional center. Quiet and always smiling, soft and full-hipped, Esther favored house dresses and plain, comfortable shoes, and rarely adorned herself with jewelry and accessories or makeup. Esther saw the management team that Rich had put in place as family, calling them "my boys." Respectfully, they called her Mrs. Snyder, and felt a genuine warmth and affection for her as well.

While the death of her husband was devastating, those that knew Esther described her as unrelentingly positive. Rather than turn sullen, she turned to her faith in God and focused on the company that she had co-founded with Harry. Each morning, Esther drove the nine miles to In-N-Out's Baldwin Park offices, arriving faithfully by 7:30 a.m., her brown leather briefcase always by her side. There she handled the chain's payroll and accounting, personally signing every check by hand, a practice that didn't end until around 1988, when the chain had grown to nearly fifteen hundred associates. At heart she was a bookkeeper; she knew every invoice and price point, and she monitored financial developments like a hawk. Even as the company grew—and the mail with it—she still opened and read each letter. Her sweet, grandmotherly demeanor notwithstanding, Esther remained razor sharp. She sat quietly in meetings, listening intently. She spoke up rarely, and when she did, everyone listened. Although she would have preferred to leave the chain largely as it was when Harry was alive, Esther unequivocally supported Rich.

With Rich running the day-to-day operations, Esther began to focus more time on her charitable work. Her dedication was rewarded when, in 1990, Baldwin Park named its new community center after her. On weekends, she regularly drove to Costa Mesa to attend services at the Calvary Chapel. Her life was a full one. Active in the Rotary Club and the California Restaurant Association as well as the company's Child Abuse Fund, Esther never remarried. When asked why, her nephew Joe Stannard, the son of Esther's sister Virginia, and a financial adviser living in Springfield, Illinois, replied, "There was only one Harry."

Her friendship with Margaret and Carl Karcher was an enduring source of support for Esther. After Harry died, the three remained in close touch. "We went to dinner twice a year," Karcher proudly explained. While the Snyders and the Karchers had earned more money than they could have ever imagined and could easily have dined at any of Los Angeles's finest eateries, they chose instead to alternate their dinner meetings between In-N-Out and Carl's Jr. Philanthropically minded, the two families supported each other's charities. One of the Karcher daughters, Barbara Wall, recalled that her parents "could always rely on [Esther]. She always came through with any personal request of dad's."

After Harry's death, Esther often sought out Carl Karcher's counsel. In fact, as In-N-Out Burger continued to grow and began to open up new stores in Carl's Jr. territory, Karcher and his wife remained uncommonly generous. "I always called the Snyders colleagues, not competitors," explained Karcher. "I joke that that was a mistake with the success they became."

CHAPTER 15

In the fall of 1988, following the opening of the Thousand Palms store in Riverside County (the chain's fiftieth outlet), Rich came up with big plans to mark another milestone: In-N-Out Burger's fortieth anniversary. Without a doubt, it was a significant event, and Rich intended to celebrate it in high style. He hired a specially outfitted passenger train just for the occasion. The Snyders invited a number of associates and guests to the private onboard fete. The anniversary party took off from Anaheim Station on October 22 and then traveled down to San Diego by rail. It was forty years ago to the day that Harry and Esther Snyder had opened their first drive-through in Baldwin Park.

As the train pulled out of the station with a jerky hiss, Esther and Rich came out and stood under a red and white In-N-Out awning at the edge of the caboose. An audience of spectators and reporters gathered on the platform below to see the party off; Esther and Rich smiled broadly and waved to the crowd below. The pair looked to be in great spirits. For Esther, it was an especially nostalgic day. San Diego was a city filled with memories—it was there that some forty-five years earlier Esther, a navy WAVES, had been stationed at the Naval Hospital.

The trip to San Diego wasn't purely a celebration for Rich. As In-N-Out's revelers made the return trip north, he got off the train to take care of some business. He had set San Diego in his sights for the chain's expansion plans. At the time, In-N-Out Burger was negotiating over a couple of sites and he decided to use the trip to survey the area. His goal was to open the first San Diego–area In-N-Out by mid-1989.

Just north of California's border with Mexico and south of Orange Country, San Diego, 122 miles from Baldwin Park, was the farthest point outside of In-N-Out's core radius. It marked Rich's most ambitious geographical push to date. "San Diego is a totally new market area for us," he acknowledged excitedly on the chain's anniversary. "So it's a big move." The second-largest city in California (after Los Angeles), San Diego had a growing population with a diverse economy and considerable affluence. It was home to miles of beaches, sixteen military facilities, fifteen colleges and universities, and a flourishing tourist industry. Perhaps more important, scores of residents had more than a passing familiarity with In-N-Out, and they had long bemoaned the absence of the beloved chain from their own backyard. For years, many drove for an hour or two up to Orange and Los Angeles counties for a Double-Double fix. As far as San Diego was concerned, a new In-N-Out couldn't open up fast enough.

It took another year and half following In-N-Out's fortieth anniversary before a San Diego drive-through became a reality. Store number fifty-seven was located in the neighborhood of Lemon Grove (just off of Highway 94). Besides being the first In-N-Out in the San Diego-area, Lemon Grove was reportedly the first store in the chain's fold to sell 1 million hamburgers in a single year. Before the year was up, Rich launched a second unit in the northern San Diego suburb of Vista, store number sixty-one.

Certainly, nobody could accuse the chain of oversaturation; the Vista outlet was located about forty-three miles from Lemon Grove. It proved equally popular, reportedly setting an opening day record by selling over five thousand burgers. Lemon Grove's tally wouldn't be broken for another nine years, until store number 137 in Redding, California, sold an estimated seven thousand–plus burgers during its own opening.

It was during this time, as Rich pushed deeper and further into new markets, that a debate erupted inside the chain's Baldwin Park headquarters about the direction of the company's future growth. By 1990, In-N-Out had grown to sixty-four stores, becoming a chain about three and half times the size it had been when Harry had died. And the chain's revenue, an estimated $73 million, continued on an upward trajectory, growing at an average rate of about 15 percent annually. Still, industry-wise, In-N-Out remained a small player. By contrast, at least in terms of size, McDonald's had about 8,576 domestic restaurants while Burger King had opened 5,468. Wendy's had reached 3,436 stores, and even Sonic Drive-Ins had passed the 1,000-unit mark. The argument in Baldwin Park centered on the nature of its continued growth; should it be slow and narrowly focused, or slow and widely spread?

While most fast-food chains deployed a deep penetration policy, opening up in every conceivable high-traffic spot—blanketing intersection corners as well as malls, locations near schools, urban centers, and airports with multiple stores in a single location—In-N-Out had remained incredibly selective about where it placed its units. Originally its drive-throughs were established in suburban and outlying neighborhoods, where the land was cheaper and could be purchased outright. As real estate prices in California continued to soar, Rich continued to follow this strategy. For instance, Lemon Grove was chosen over the more expensive tourist corridor along the Interstate 5. As Rich had told *Forbes*, "Why pay $2 million for a property if I'm going to sell the same amount of burgers as I would on the property I pay $500,000 for?"

Unlike nearly every one of its competitors, In-N-Out rarely if ever put more than one drive-through in the same market. The fact that there were so few locations, spaced so far apart, only added to In-N-Out's allure. It made eating a Double-Double a conscious act, an experience. As big burger chains employed a host of consultants and even commercial satellite photography to capture the best potential locations, In-N-Out's consumers proved time and again that they would go to extraordinary lengths to eat at In-N-Out.

A key part of In-N-Out's location strategy was to find spots that would generate high volume and large turnover. For the most part, In-N-Out still catered to the car-reliant consumer. In 1989, following the San Diego push, the chain opened its fifty-sixth store in Barstow, California. An important silver mining town and busy railroad point in the Mojave Desert during the 1860s, more than a century later, Barstow was known for little more than standing at the entry point to California on the old Route 66. A small, hot, and dusty afterthought of a town, Barstow had lost much of the luster of its glory days, when it was called Calico Junction. However, it did sit on an important crossroad: the intersection between Interstate 15, Interstate 40, and State Highway 58. In the late twentieth century, Barstow had become a popular rest stop for those on their way to or from Las Vegas.

Store number fifty-six opened up on Lenwood Road, not far from the Mojave National Preserve and the U.S. Army National Training Center; its large customer base was not found among those who were staying, but among those who were just passing through. It was a smart move, and business took off. The Barstow drive-through was the first In-N-Out to be built with three grills and it had the largest indoor seating capacity of any In-N-Out to date (almost three hundred people). The line approaching the counter in Barstow was divided by metal bars to keep customers orderly and moving forward.

As Rich pushed the chain's geographical boundaries, each store was still within a radius that allowed the company to deliver its fresh beef patties and produce from the Baldwin Park commissary on a daily basis. It was obvious that the slow, controlled rollout strategy had contributed greatly to In-N-Out's mystique and popularity. Rich and his crew of relatively new professional managers were pressing for a broader expansion. At the same time, some of the top managers, including many of the old-timers who had started under Harry, believed that In-N-Out's mystique was one of the chain's most important assets. Many of those longtime managers spearheaded the argument against further growth at the expense of its mystique. Would In-N-Out be able to retain its intimate, local feeling as it expanded?

Would In-N-Out's enigmatic appeal and personal touch be sacri-
ficed by a further rollout? Could In-N-Out remain In-N-Out? It was
a question that even the largely absent Guy Snyder had considered.

Although the paucity of stores had helped to drive demand, it
also led to traffic jams and idling cars that blocked parking lot exits
and entrances, obstructed parked cars, and irritated the owners of
nearby businesses. The long lines had become the chain's unofficial
trademark. For the most part, In-N-Out's customers appeared not to
mind waiting twelve to fifteen minutes (or more) for their orders.

However, Rich had some practical issues to consider. Given the
limited number of stores, even if no new stores were opened, the
chain needed to find some way to relieve the congestion. For a time,
the company tried to alleviate the bottleneck by adding extra grills
and fryers to help speed up each store's ability to cook orders. (The
chain would later deploy associates with PDAs to help move the
drive-through lane along.) But those measures only helped so much.

Rich had learned from a series of focus groups that the mystique
of his chain only went so far. After talking about how special In-N-
Out was, the groups complained about the long lines and distance
between stores that left their fries cold by the time they arrived home.
What customers wanted were more In-N-Outs—and fast. In the end,
Rich went with his gut and decided to proceed with his rollout. In
executing his desire to open some ten new stores per year, Rich had
to reconsider certain factors of the chain's longtime growth strategy. Cautiously, he had begun moving out of the sites in peripheral
suburbs to more central locations. In 1989, In-N-Out opened store
number fifty-five in West Los Angeles on Venice Boulevard, near the
10 Freeway. Gradually, the chain opened stores in other, bigger cities and urban locales such as Hollywood (on Sunset Boulevard) and
eventually San Diego proper (near the I-5 corridor).

In another break with the Harry Snyder tradition, Rich also began to open new stores in markets where an In-N-Out was already
located—slowly, sparingly, and, for many years, not within a six-mile
radius of one another. In doing so, the company had to be ever more
vigilant in monitoring its locations. It wasn't just a highly trafficked

intersection or freeway off-ramp that made an In-N-Out a success; although In-N-Out's prices remained modest, the company found that the highest volume stores were located in higher income areas. As In-N-Out began to put more than one store in a single territory, income was a criteria that the company could not ignore.

At first, some of the managers (whose bonuses were tied to their stores' profits) were concerned that the new strategy would eat away at their business—but the company had conducted impact studies that showed the opposite to be the case. In fact, after an initial drop-off, the original store's volume not only rebounded but increased.

When Harry and Esther turned the reins over to their youngest son, Rich, In-N-Out owned its warehouse, offices, and meat department on East Virginia Avenue, in addition to all eighteen of its stores. Harry was strictly old school; if he couldn't afford to own a property outright, he didn't need it. Rich did not operate under the same conservative rubric. He wasn't afraid to go outside of the company for financing and sought a line of bank credit in order to facilitate his expansion plans.

At the start of the 1990s, there were fewer than one hundred In-N-Out stores—not enough units to generate the kind of income needed in order to open ten new stores a year. The next best option was to lease. For one, it helped ease up the cash flow. Secondly, Rich was also finding that increasingly he was up against large property developers. If he wanted to continue to operate in key locations, he had no choice but to lease from them. One of In-N-Out's first leased stores was in Norco, a rural equestrian community in northwest Riverside County where hitching posts remained a common sight. Around 1990, In-N-Out owned all but a handful of its sixty-four stores. That changed over time; eventually, the chain owned roughly 60 percent of its stores and leased the rest.

Around this time, Rich established a separate business entity called Snyder Leasing that received income based upon a percentage of the gross income generated by a number of In-N-Out stores as well as a number of commercial properties. While the company purchased the land and built the properties, the stores listed in Snyder

Leasing were deeded over to Rich, who was its primary beneficiary. Between 1991 and 1992, perhaps ten In-N-Out units were listed in Snyder Leasing's portfolio of properties; not surprisingly, they were some of the highest-performing shops in the In-N-Out fleet.

It was a potentially lucrative move. In-N-Out was classified as an S-Corporation, largely for tax purposes. S-Corporations are not taxed on profits; rather, its owners are taxed on their proportional shares of the company's profits. As In-N-Out's president, Rich's salary had a ceiling for tax reasons, and Snyder Leasing gave him another stream of revenue.

In his official capacity as executive vice president, Guy Snyder oversaw the meat department and warehouse. In actuality, he could be found at the Baldwin Park complex on average not more than two or three times a week. He was not involved in the chain's day-to-day operations. Marriage and fatherhood notwithstanding, he had yet to shake his reputation as a rebel. It was an assessment usually colored by Guy's own cycle of behavior. Impulsive, Guy might decide to fly to Cancun one week, taking a group of friends with him—or he might spend the day riding dune buggies. Once, a friend recalled, Guy went to another fast-food restaurant and ordered one of everything on the menu just to see the reaction of the young cashier.

Increasingly, Guy and Rich were not on the same page. In the years that followed their father's death, Guy distanced himself further from In-N-Out's daily workings and from the Baldwin Park corporate offices. More than anything, Guy seemed to follow his own interests. Around 1988, he opened the Flying Dutchman Tobacco Hold in Glendora, a store that sold exotic blends of tobacco and pipes from all over the world. More of a hobby than a serious business venture, the Hold closed for good after only about two years. He spent a great deal of time on his 170-acre ranch up in Shasta County, and soon the family made the Flying Dutchman their main residence.

Drag racing was Guy's primary vocation, and during the late 1980s, he stepped up his already considerable involvement in the sport. He had amassed about half a dozen racers including a green Dodge Charger, a Blue Dodge Dart, and a blue 1941 Willys—but it was his 1986 IROC Chevy Camaro that he raced seriously. When he wasn't racing on the official circuit, Guy, who raced sportsman's class, frequently rented out drag strips in California and Arizona for his own personal use. He'd call up his crew and they would haul his trailer to the Famoso Raceway in Bakersfield or the Firebird Raceway in Phoenix, where he ran his own races.

In 1984, the same year that Rich launched In-N-Out University, Guy convinced his mother and brother to sponsor a professional NHRA team. It was a particular desire of his to see the In-N-Out logo displayed on a winning dragster. The family started small, initially paying a few thousand dollars to partially sponsor the Over the Hill Gang's Funny Car. The Over the Hill Gang was a group of five friends from the San Gabriel Valley who got involved with drag racing in the 1970s. "We could afford to partner up and build the car," explained Joe McCaron of the Gang team, "but we didn't have enough money to race it." (Financing a successful NHRA team could cost more than a million dollars.)

In-N-Out's sponsorship increased incrementally over the years from partial sponsorship to full sponsorship, underwriting about three races a year for the team. However, by 1987, In-N-Out's management had become concerned about the money they were shelling out for the team, particularly because it hadn't won a major race. As luck would have it, that same year the team qualified and landed on TV, In-N-Out was just about to pull its funding. In 1988, the Over the Hill Gang won the Winternationals in Pomona. The Snyders and all of In-N-Out's associates were thrilled. The winning Funny Car was pulled through Covina's Christmas Parade and displayed at In-N-Out's Christmas celebration. At the party, the children of associates were allowed to play with the car, pretending to race, sitting behind the wheel.

In 1989, Guy hired Earl Wade, the longtime, highly regarded tuner for Don Nicholson, to build his engines. Wade had been something of a legend on the drag racing circuit from its earliest days, and the pair had known each other for about fifteen years. Guy had hoped to take his racing to a professional level and wanted Wade to help him. The two men set up a meeting at Earl Wade Racing, the tuner's shop in Monrovia. "Here was a man with a lot of money that can do anything he wants," was how Wade described Guy Snyder. "Money was never an object. He sure didn't lack in that department."

In fact, the Snyders kept a tight rein on Guy's spending. Drag racing might have been Guy's hobby, but he didn't hold the purse strings. "Mama and Rich, they oversaw everything that Guy did," recalled Wade, who said that the Snyders' main concern was for Guy's safety. Rich saw this as an opportunity to encourage his brother. He also viewed the venture as something of an investment in advertising for In-N-Out. The family came up with a private plan, and the Snyders agreed to fund Guy's racing. Anytime Guy needed money for the venture, he had to go through Rich. It was Esther who signed the checks. The annual budget discussions were often the source of blowouts between the brothers.

Rich's support was indicative of the brothers' complicated relationship. As one of Guy's friends put it, "They loved each other, they fought a lot, and they hurt, but they both had big hearts and cared about the underdog. I think Rich really wanted Guy to be happy, and Guy loved racing, and Rich wanted to help him."

One of the people whom Guy liked to bring along on the road to his races was his friend Tom Wright. Wright, the nephew of Guy's wife, Lynda, reportedly worked for the FBI in the late 1970s and later in security for large Southern California–based aerospace companies. In 1987, Guy brought him to work at In-N-Out as an asset protection investigator. The asset protection department was concerned with protecting In-N-Out against theft and robberies at the chain. According to Tom's wife, Dale, the two had become quite close. Dale worked as Lynda's personal assistant, picking up Lynsi from school and doing errands. Frequently, Guy insisted that Tom drop everything and accompany him to his races. It was a situation that irritated Rich and In-N-Out top brass because it meant that Wright would be absent for stretches of time, neglecting his responsibilities.

Once, while Guy was on the road to Arizona (where he planned to test some of his racers), Phil West, In-N-Out's number two executive, called at Rich's behest to complain that Tom was needed to deal with some unfinished security-related matters. According to an individual familiar with the episode, "Guy barked back, 'tough shit.' It really pissed off Rich." Guy remained unconcerned; he took a myopic view of the situation and seemed to shrug off any problems he might cause for the executives in Baldwin Park. In his mind, racing came first, and whatever In-N-Out–related work Tom could fit in around that was just fine with him.

Back in Baldwin Park, it had become obvious that Rich was the Snyder brother truly running In-N-Out Burger. Rich tried to give Guy small tasks, but more often than not, Guy let his brother down, and Rich did not trust him with larger duties. When the chain received press attention, it was Rich who was quoted and photographed; Guy remained deep in the background. While Rich carried the considerable burden of running In-N-Out, his older brother was off following his own sense of priorities.

Guy came to resent his brother and the position that left him dependent on Rich for access to funds and under his authority at In-N-Out. The dynamic of their relationship left Rich both angry and sad. Underneath the acrimony, Rich did want Guy to succeed and be happy. He felt particularly helpless when it came to Guy's struggles with drugs. At times, Rich was brought to tears over the situation.

Still, the Snyder brothers came together on numerous family and business occasions. They attended the annual company fishing trip for In-N-Out executive managers at Madison River in Ennis, Montana. Although it was an opportunity for the men to have a few days to be boys, according to one account, during the weeklong trip the pair regarded each other cordially but warily, sticking pretty much to their own entourages. As one friend described the scene, "We all stayed in the same place, but it was like two teams. Richie had his guys close to him, and Guy would take those guys close to him."

Helping to maintain a buffer between the two brothers was a man named Rick Plate. Plate had known and worked for the Snyders for years. Something of a jack-of-all-trades, he served as a right hand to Rich and Guy as well as acting as their go-between. Given a personal services account, Plate was dispatched to purchase items the brothers needed—including houses. Often he was tasked with handling details surrounding Guy's racing or Rich's hot air balloon hobby. Usually he was asked to be at two places at once, babysitting Guy and handling the Snyders' needs, often just keeping them out of each other's way.

Rich wanted to take his business to the next level. When it came to In-N-Out Burger, Guy was rather conservative-minded; like Harry and Esther, he wanted to keep things much as they always had been. He was concerned that In-N-Out might lose its intimate feeling as it got bigger. Once, when Guy had heard that Rich was considering the possibility of introducing a chicken sandwich, he became incensed. "He went ballistic," recalled a friend. "He was not going to go for that at all."

The two fought often, their rows sometimes erupting into shouting matches that reverberated throughout the Baldwin Park headquarters from the executive suite on the second floor. "They fought over just about anything," recalled the friend. "It didn't have to be something big, other than the fact that they pretty much didn't agree on anything."

Their head-butting reached numerous climaxes—perhaps a standoff was inevitable. Guy continued to struggle with drugs and he could be volatile. It was something Guy's friends on the racing circuit could hardly ignore. There would be periods when Guy was at the track and long stretches when he would simply disappear. Increasingly, Guy and his private troubles were becoming a company liability. He was said to deploy some of his assistants to help him get prescription painkillers from a series of doctors. To a large degree, Rich felt like his hand was forced. While he viewed his older brother as radioactive, privately, the whole situation tore him up.

In one bitter incident, their rocky relationship blew up over a T-shirt.

For years, the company's signature T-shirts had been extremely popular. In fact, by the summer of 1989, the chain was selling twelve thousand T-shirts per month, spurring Rich to launch a mail order catalog of In-N-Out merchandise that later grew into a retail shop. Called the Company Store and initially located at the headquarters complex in Baldwin Park, its inventory grew to include such logo items as mugs, key chains, magnets, and beach towels. In-N-Out's T-shirts had come to be considered a souvenir of "real" California. They could be found all over the country and even on the streets of foreign cities. The shirts were designed with airbrushed images intended to evoke memories of a simpler period, when the chain first began, and before fast food took over every street corner. Most of the T-shirts featured cars from the Snyder brothers' own collection of classics and other telltale symbols that signified both In-N-Out and Southern California: palm trees, hamburgers, the beach, and the famous yellow In-N-Out boomerang arrow.

Then Guy came up with his own design for a shirt. His featured a sexy girl straddling a hamburger, and he had a couple hundred of them manufactured. "I think he knew when he made them that they weren't going to sell them at the store level," recalled a friend. In all likelihood, it was his way of taunting the straightlaced Rich. If that was the intention, it worked; Rich was infuriated. An unpleasant conversation followed—the fallout was swift. The T-shirts remained packed in their cardboard boxes.

It wasn't long before Rich had reached the end of his tether; he wanted to run the company without any interference.

The formal titles given to the brothers reflected not just Harry Snyder's last wishes, but the roles that they had been assigned while growing up: Rich was the reserved and responsible natural-born entrepreneur; Guy was the liberal, partying, wild-child older brother. They were the unequal heirs of a family business, and their positions reflected that. As the younger brother who had worked at In-N-Out Burger continuously since he was sixteen years old, Rich held the title of president. He ran the show and he was Guy's boss. In a move that would have long-ranging repercussions before the decade was out, the family moved to legally formalize what was already a fait accompli.

On January 31, 1989, the Snyders established an irrevocable family trust: the Esther L. Snyder Trust. The trust was set up to ensure that a majority ownership of the company remained in the hands of direct lineal heirs of Harry and Esther Snyder. At the time, Rich was still single and Guy, who had been married for about eight years, had a seven-year-old daughter. A separate trust, the Lynsi L. Snyder Trust, was set up for her.

Esther, who held a controlling interest in the company, transferred to the trust a majority of the shares of capital stock of In-N-Out for the benefit of her sons. Esther declared her two main purposes in establishing the legal instrument: first, "to make provision for my two sons and other lineal descendants, including my granddaughter Lynsi," and second, "to enable the stock ownership of my closely held corporation to be transferred from one generation to the next such that the business of the corporation could continue to grow and prosper."

According to court documents later filed, the Esther L. Snyder Trust was made up of 44,147 shares constituting 65.85 percent of the issued and outstanding shares of capital stock, a controlling interest in the company. The Lynsi Snyder Trust held 4,370 shares of In-N-Out stock, 6.52 percent of the total issued and outstanding shares of capital stock of the company.* The instrument was typical for a closely held family company that intended that shares would be gifted or willed to the next generation.

However, the Snyder trusts also made several specific points perfectly clear. It was the Snyders' expressed intention to keep the company closely held within the Snyder family, following a direct line of

---

* The trusts' share amounts are as of court filings in 2005 and 2006.

blood descendants. Underscoring this intention was a provision specifically addressed to exclude Guy's stepchildren. "For purposes of
this instrument, Traci Lynette Taylor and Terri Louise Perkins shall
not be considered issue of Esther or Guy," proclaimed the trust in no
uncertain terms.

A second, perhaps more significant provision of the trusts concerned the distribution of their assets. While Esther created her trust
for the benefit of both her sons, it was not set up for them to inherit equally. The assets of the Esther L. Snyder Trust were divided
into two disproportionate parts. Rich was to receive 89.8224 percent
while Guy was to receive 10.1776 percent to be distributed in 1996,
seven years after the trust was set up, or in the event that Esther died
prior to 1996. As the trust itself controlled the majority of company
stock, Rich would maintain majority ownership of In-N-Out and his
brother Guy would not. Although Guy didn't grouse about the situation publicly, one of his friends put it this way: "He was the oldest
son. It had to rub him the wrong way."

At the time, the trust was structured to reflect the reality of the
business and the family, the wishes of Esther Snyder, and her expectations for the future. That reality, however, was about to change—
and nobody saw it coming.

CHAPTER 16

Rich Snyder liked to think that success could be achieved through honest dealings. Even during troubled times, he believed in fair play. For him, a man's word was his bond. It was more than just an aphorism. Rich lived that philosophy, and he expected the people he surrounded himself with, including business partners, to act similarly. As In-N-Out Burger was marching toward opening one hundred stores, its reputation growing with each new drive-through, Rich found himself in a situation that tested those beliefs.

If anything, the Snyder family had been unfailingly loyal. Many of In-N-Out's vendor and supplier relationships stretched back decades, to the time of Harry's first two-way speaker box. Begun with a handshake and held together through goodwill, they represented the importance of personal relationships to the Snyders' business philosophy. One of In-N-Out's longtime vendors was PepsiCo. Over the years, the family had been approached by the Coca-Cola Company, but they politely refused to make the switch—in large part out of loyalty to PepsiCo and their relationship with that company.

Then, in 1990, PepsiCo purchased a six-year-old, Michigan-based double drive-through hamburger chain called Hot 'n Now. PepsiCo was in an acquisitive mood, having already purchased Taco Bell, Pizza Hut, and California Pizza Kitchen, folding them into its growing restaurant subsidiary. The majority of Hot 'n Now outlets were drive-through only—the buildings had slanted roofs and a yellow lightning bolt logo. The chain featured a simple menu of hamburgers, french fries, and soft drinks. Rich was concerned about the chain's similarity to In-N-Out Burger and possible encroachment on his territory, and he shared his concerns with PepsiCo. The discussions reportedly went high up the management ladder. Rich was assured that the company would not expand Hot 'n Now to the West Coast. However, a year later, as he was scouting locations in Fresno and Las Vegas, Rich found himself competing for prime spots with none other than PepsiCo's Hot 'n Now.

Rich felt betrayed. In-N-Out Burger had ties to PepsiCo going back to his parents' first days in the business. The way Rich saw it, In-N-Out Burger had a longtime bond with PepsiCo; by going back on their word, the company had broken that bond. Rich proceeded cautiously. He did a kind of quiet beta testing, replacing Pepsi-Cola with Coca-Cola in ten stores to see how the customers would react. The response was positive, and in 1992 Rich made the switch a permanent one, substituting Coke for Pepsi chain-wide.*

As Rich dissolved In-N-Out's partnership with PepsiCo, for the first time, the chain stepped outside of California. In 1992, he opened store number eighty in Las Vegas on Sahara Avenue just west of the Interstate 15. To mark the burger chain's move, the first of its kind in the company's entire history, Rich had an In-N-Out semitruck parked on the state line between California and Nevada near Whiskey Pete's Hotel and Casino. Once there, a thousand-foot rope was tied to the semi. When Rich gave the signal, a group of managers tugged

---

* As it turned out, there was no happy ending for Hot 'n Now. During the next seven years, the chain expanded from seventy-seven to two hundred stores before it began to fail; eventually, all but seventy-five locations closed, and PepsiCo sold the chain to an entrepreneur named Ron Davis. According to *Nation's Restaurant News*, PepsiCo recorded a $103 million charge against its restaurant subsidiary earnings in fiscal 1995 that was almost entirely attributed "to the write-off of Hot 'n Now." By 2005, the chain had dwindled to fourteen stores and Davis sold it to STEN Corporation for $175,000 in bankruptcy proceedings.

the rope and pulled the truck across the state line into Nevada. Rich, who always liked to make a big splash, also hired an Elvis Presley impersonator for the occasion. "Elvis" flew into the parking lot, carried aloft by a hang glider bearing the In-N-Out logo. Store number eighty was the flagship of the eight shops eventually established in Sin City. In-N-Out's second Las Vegas drive-through opened less than a year after the first; located on Dean Martin Boulevard, store number eighty-six had what was reportedly the world's largest neon fast-food restaurant sign. Las Vegas was more than 250 miles from Baldwin Park, and before long, the chain established a Vegas-based distribution center. A satellite of the Baldwin Park Commissary, it allowed In-N-Out to maintain its quality standards by delivering fresh ingredients to its stores daily. Although the chain maintained its typical silence, In-N-Out fans, industry watchers, and other hopefuls saw the chain's entry into Las Vegas as a sign that In-N-Out Burger finally intended to expand its geography.

That hope became something of a mantra among In-N-Out's far-flung and extremely loyal fan base. By the early 1990s, the host and staffers at *Late Night with David Letterman* were known to frequent In-N-Out whenever they were in California. In fact, the group usually made an In-N-Out run part of their annual trek to the Emmy Awards, stopping at the burger joint sometimes still dressed in their tuxedos and gowns on the way to the airport. As one staffer described the annual celebration, "For those of us who don't always go to the awards show, the Monday morning water cooler conversation usually goes something like this: 'How were the Emmy Awards?' 'Great! We stopped at In-N-Out Burger on the way back.'"

The new decade was off to a splendid start. For Rich, it appeared that his business success was now matched by success in his personal life. Two months before his fortieth birthday, the timid bachelor married for the first time on May 2, 1992. A friend had introduced Rich to his new bride, a twenty-six-year-old sales associate at a Newport Beach equipment leasing company. Sloe-eyed and willowy with chestnut hair, Christina Bradley had a young daughter named Siobhan. The new couple shared a strong Christian faith and a commitment to helping disadvantaged and abused children. "He was just larger than life," was how his new wife later described him. "He had a magnetism that just drew you to him."

The couple was married in a small green church in Maui. Their fun-filled, 1950s-themed reception was held at the Grand Wailea resort. Although it was small, no expense was spared. Rich paid to fly eighty-two of the couple's friends and family to Hawaii (Guy and his wife, Lynda, were conspicuously absent from the wedding party). He also had his classic 1957 white Cadillac convertible shipped across the Pacific for the occasion. Rich's friend and spiritual adviser, Pastor Chuck Smith Jr. of the Cavalry Chapel (son of the church's founder), presided over the ceremony.

In June, just weeks after the Snyders' Hawaiian nuptials, the newly-weds received an engraved invitation from President George H. W. Bush and his wife, Barbara, to join them at a state dinner at the White House in honor of Russian president Boris Yeltsin. Although Rich had attended Bush's inaugural festivities in 1989, he could barely contain his joy. In the final months of the Bush administration, Rich was finally able to check off the two-year goal that he had first set for himself twelve years earlier in 1980—he had received a White House invitation.

Arriving at the White House, Rich and Christina Snyder found themselves in the company of such notables as Kenneth T. Derr, the chairman and CEO of Chevron Corp.; Louis V. Gerstner, then the chairman and CEO of RJR Nabisco, and his wife, Robin; Federal Reserve chairman Alan Greenspan; and a Stanford University professor of political science named Condoleezza Rice. After dining on a menu of caviar, roast loin of veal, and caramel mousse in the elegant State Dining Room, the Snyders (along with the president's other guests) moved to the East Room to listen to soprano Carol Vaness.

It was the best party that Rich Snyder had ever attended. As he excitedly told Karen De Witt, a *New York Times* reporter who covered the evening, "I love history, I love our country, and it was all there. The classiest thing I've ever been to. And it wasn't stuffy, either." This time it was Rich Snyder who felt as if he had won the golden ticket to enter Willy Wonka's chocolate factory. In-N-Out's president spent the entire evening slack-jawed, pinching himself, overcome by the opportunity to chat up Defense Secretary Richard Cheney, General Colin Powell, and Vice President Dan Quayle, who a year earlier made a well-publicized stop at the Kearny Mesa In-N-Out. The absolute pinnacle of the evening came at about 11:00 p.m. As Yeltsin and his wife, Naina, said their good-byes, President Bush motioned to the Snyders to join him and Mrs. Bush on the dance floor, where the two couples two-stepped to "Shall We Dance." "There we were, dancing with the President and Mrs. Bush, the only ones on the dance floor," Rich marveled.

At the end of the evening, a giddy Rich pocketed his nametag, the evening's menu, and even the table placard to keep as souvenirs—much to the amazement of his wife, who was already slightly embarrassed by the number of photographs he had taken during the dignitary processional on the White House lawn. Back home in California, Rich recorded his entire recollection of the evening and had his secretary type up the transcript. "So that in years to come," he explained, "I won't forget a single detail."

Following Rich and Christina's wedding, the newlyweds moved into his $3.5 million waterfront home on Bayshores, one of the most exclusive enclaves on Newport Beach's Lido Isle, frequently referred to as California's Rivera. A narrow strip of land tethered to Newport Beach by a short bridge, Lido Isle was transformed into a residential playground during the 1920s by a millionaire developer by the name of W. K. Parkinson, who had made his fortune in oil.

Rich, who had been living in Newport Beach when he met Christina, had begun the process of remodeling his house in anticipation of starting a family. The pair planned to name their firstborn son Harry. Christina suggested that they first get a dog as a kind of test run; Rich, however, wasn't a dog lover, and he wasn't immediately sold on his wife's idea.

Devoted to his mother, Esther, Rich also purchased a second home for her so that she could live close by. In an area of obvious wealth, Rich delighted his neighbors during their frequent block parties when he insisted on bringing an In-N-Out trailer and supplying all of the hamburgers. "He liked being happy," said his friend and next door neighbor Bob Longpre, who owned a Lexus dealership in nearby Westminster. "With Rich, what you saw is what you got." Described as a kind of wealthy everyman, Rich could be found in the local grocery store or working on his pride and joy—a twenty-one-foot custom Duffy Electric Boat Company boat christened *Watts-A-Luck* that was docked in front of his home. Duffy also manufactured electric boats for Walt Disney; *Watts-A-Luck*, outfitted with gold-plated parts, was reportedly one of the fanciest of its kind ever built.

Not long after his wedding, Rich began strategizing for both his and In-N-Out's future. With his personal life settled in Newport Beach, Rich decided to move In-N-Out's corporate headquarters closer to his new home from its longtime seat in Baldwin Park to the city of Irvine. It was a bold move on his part; for one thing, it signaled a huge break in continuity. Moreover, Baldwin Park had unmistakably influenced the company and its culture. In-N-Out Burger had grown up in Baldwin Park, and Baldwin Park was considered by many to be the heart of the company. Irvine—in Orange County—was less than an hour's drive and about forty-five miles away from Baldwin Park. In reality, however, the distance between the two cities was much greater.

The Baldwin Park of the first In-N-Out had disappeared. The San Gabriel Valley had long since been bulldozed, divided, and finally subdivided into a tedious mosaic of mini-malls, vacant lots, and tract homes. The Arcadian image that had lured millions to the area remained largely on those postcards and citrus crate labels that once "advertised" the Valley's scenic wonderland of oranges and sunshine. A working-class suburb whose rough edges had become increasingly part of the city's center, it was facing many of the problems of Southern California's postwar boom communities: increasing crime rates, gangs, shifting demographics, and an eroding manufacturing economic base.

The last vestiges of Baldwin Park's bucolic charm—such as the cast-iron smudge pots used to protect the orange groves from frost in the winter—belonged to history now. The dairy farms and chicken ranches were long gone. Even the famous Vias turkey ranch had moved to Apple Valley seventy-five miles away, where it was converted into an ostrich ranch.

Affluent and manicured, Irvine was one of the nation's largest planned communities, and it contrasted greatly with the blue-collar scruff of the city that Baldwin Park had become in the years since the Snyders had first arrived. During the 1960s, the Irvine Company, a developer of planned suburban cities, designed the entire city (incorporated in 1971) around the newly built campus of the University of California. Long before the Walt Disney Company developed the city of Celebration in Florida, every detail of residential and industrial Irvine was planned, right down to the construction of its bike lanes and man-made lakes. A post-post-boom city, Irvine's two dozen townships, spread over fifty-five square miles, were each separated by six-lane streets. A quilt of commercial districts bordered the central villages, and the villages were placed sequentially along two parallel main streets that met at the university campus.

In-N-Out Burger planned to take up residence on the top two floors of the ten-story University Tower on Campus Drive on the edge of the university's grounds, leasing the remaining eight floors to other businesses. A modern glass box of a building, it had few of the charms of the Spanish-style headquarters that Rich had built some ten years earlier. (So as not to put off vendors and associates, Rich designed the ninth-floor reception area to resemble an In-N-Out walk-up window.) As one insider noted, the chain's new executive offices looked like the set of the television series *L.A. Law*, which was popular at the time. Rich loved the building, and it signaled the beginning of a new chapter in In-N-Out's history.

Irvine's mayor, Sally Anne Sheridan (who was also a real estate agent and former member of the city council), was thrilled that In-N-Out was setting up its new corporate headquarters in her city. "They called me to talk to me about moving the company down here and asked if the city would welcome them," she remembered. "We reassured them that we would do everything to help. They were very good corporate friends. I was really impressed with the company."

News of the move came as something of a surprise to the residents of Baldwin Park. Word reached the community through real estate circles only a few months before the planned move, when escrow opened on the building. "A lot of us were surprised and disappointed," recalled Bob Benbow. "In-N-Out had played an important part in the development of Baldwin Park, and it was a point of pride here."

Indeed, many felt that In-N-Out had put Baldwin Park on the map much in the same way that Disneyland had done for Anaheim (albeit on a much smaller scale). Certainly, the chain distinguished the city from the numerous postwar boom communities swallowed up in Southern California's growing suburban sprawl, connected to other suburbs by the vast network of freeways. More than just a successful company, In-N-Out, as Bob Benbow explained, had been "an important symbol for Baldwin Park." The two entities were intertwined. In economic terms, the chain contributed immeasurably to the city's bottom line. In time, In-N-Out became one of Baldwin Park's largest employers. It had also become one of the city's top three taxpayers (slipping into fifth place only after Kaiser Permanente Medical Center moved to town).

Although Baldwin Park had grown tremendously since its rural heyday, it still had something of a small town feel about it, especially among its longtime residents. There was much speculation about the transfer. In-N-Out had plucked the nondescript suburb from obscurity, injecting it with its corporate culture. Baldwin Park had become a kind of genial company town and the company was hamburgers.

A number of explanations and theories concerning the move made the rounds. In one version, the transfer was prompted by Rich's desire to raise a family in the more prosperous and comfortable Orange County. Another version of events held that Rich harbored a grudge against the city following the city council's rejection of a proposal in 1990 to rename East Virginia Avenue "Hamburger Place" in honor of In-N-Out Burger. Actually, the opposition came mainly from the local businesses that were also located on East Virginia Avenue. Some 119 businesses (including a lumber firm and a sheet metal house) circulated a petition resisting the change; it received fifty signatures. As Brent Taylor, president of Award Metals Inc. (one of the largest sheet metal manufacturers in California), told the *San Gabriel Valley Tribune*: "A name like Hamburger (Place) is totally inappropriate and embarrassing for the other 118 businesses on the street who do not sell hamburgers." * Mayor Sheridan dismissed all of the talk as uninformed gossip. "It was a good strategic move," she recalled. "They wanted to relocate down here from Los Angeles because Rich Snyder had a house down here. He wanted to move the whole company down so that he wouldn't have to go back and forth into L.A. everyday."

As usual, the company offered little in the way of a formal explanation, and so rumor filled the space where a public announcement might have gone. More than anything, Baldwin Park's longtime residents were left with the feeling that their favored son had finally grown up and was now moving away.

It was decided that the warehouse, distribution center, and the trucking facilities would remain in Baldwin Park. The decision offered a small measure of comfort to many who believed that the facilities were the heart and soul of In-N-Out Burger.

---

* A year later, in 1991, the street was officially renamed Hamburger Lane.

CHAPTER 17

December 15, 1993, was a particularly busy day for Rich Snyder. Just days before Christmas, his schedule was full, leaving him little time to slow down before the holidays hit full swing. For one, he was overseeing the final move of company headquarters to Irvine, expected to take place in two months. On that Wednesday, Rich was going to attend the grand opening of In-N-Out Burger's new Fresno store, the chain's ninety-third. During the day, he and a group of In-N-Out executives including his mother, Esther, executive vice president and aide-de-camp Phil West, and Jack Sims, a public relations executive who also acted as a company consultant, also examined a number of potential store locations.

In-N-Out was opening about ten new stores each year, and despite the recession of the early 1990s, the company was averaging 15 percent annual growth. The rollout was increasing at such a rate that the company had leased a private jet in part to ferry executives scouting new locations and checking on existing ones. Commuting by air to keep tabs on the growing chain had become so integral to In-N-Out Burger's procedures that in December, the chain was granted its application for a six-month extension to build a helipad at the company's Baldwin Park complex.

The previous August, not long after opening his seventy-fifth store, Rich gave an interview to the *San Gabriel Valley Tribune* in which he said that despite the company's growth spurt, he had no interest in competing with the fast-food industry giants. Reiterating his commitment to keeping the chain private and family-owned, he once again quashed rumors of an imminent IPO. "I think it would be too difficult to maintain quality control," he explained. "I like the fact that I can visit all of our locations and they all know me. It's kind of like what they say about farming—the best fertilizer there is in the field is the farmer's footprints."

Rich strode into the Fresno store on South Second Street. He was wearing one of his trademark suits, a tie, and—as usual—a big smile. At forty-one, he still had a boyish look about him. According to his friends, he appeared to have a real tangible sense of contentment about him. Usually upbeat, Rich seemed to be in particularly good spirits. Of late, he had been even more vocal and demonstrative with his friends and family, sliding his arm around them and expressing his appreciation for their friendship, letting them know how much he cared. He also began telling his intimates that he was "at peace with the Lord." To more than one, he said, "If I don't see you again—I love you."

The Fresno store opening went without a hitch. Rich and Esther Snyder, Phil West, Jack Sims, and a team from Baldwin Park spent time chatting with the new store's managers and associates. As had become standard, a huge crowd of enthusiastic fans had formed outside the store before it opened and the team watched as the lines swelled. It was a sight that always seemed to astonish Rich : "He told me one time that he didn't know why In-N-Out was so successful," recalled his friend and neighbor Bob Longpre. "He was just as surprised as everybody else. They would open up a store, and there was a line of people waiting to buy hamburgers."

After the Fresno opening, Rich boarded a chartered Westwind 1124A jet with Sims and his mother. In-N-Out Burger had a policy prohibiting its top executives from traveling together, so West, the company's number two, headed out to the airport to catch his own commercial flight. A number of other executives who had accompanied the group to the Fresno opening had dispersed, making their own arrangements to return to Baldwin Park or other points in the field. However, when West discovered that his own flight was going to be delayed, he circled back to the private airfield. It had been a long day. Although anxious to get home, West inexplicably joined Rich on the chartered jet. Another executive, In-N-Out's vice president of operations Bob Williams, also boarded the plane. Esther Snyder wasn't feeling well and asked to deplane as soon as possible. The jet flew to Brackett Field in La Verne (east of Los Angeles) where she got off, as did Williams. Taking off again, the chartered plane flew south toward Santa Ana.

Less than thirty minutes after taking off, trouble struck. On approach to the John Wayne Airport, the Westwind found itself trailing in the flight path of a United Airlines Boeing 757. When they were eight to ten miles from the airport, air traffic controllers at John Wayne warned the Westwind to slow down, as it was gaining on the traffic up ahead. When the Westwind was five miles from the airport, another traffic controller gave the plane a second warning to reduce its speed, suggesting to pilot John O. McDaniel and his copilot Stephen R. Barkin that they make an S-turn if necessary. Although the small jet was gaining, flying thirty knots faster than the United Airlines flight, the two pilots didn't seem too concerned. "Yeah it's close," McDaniel said to Barkin, "but I think we'll be OK."

At about 5:30 p.m., the Westwind had descended to 1,100 feet. Just one minute from the airport, the small jet was also two miles behind the 757. Suddenly the plane became ensnared in the commercial liner's wake turbulence. The pilots lost control. Within seconds, the plane rolled 360 degrees, plummeting to the ground at a 45 degree angle before crashing near the Santa Ana Auto Mall. The impact spewed charred and twisted metal in its wake. Rich Snyder, Phil West, Jack Sims, and the two pilots were all killed instantly.

A year earlier, a similar crash occurred in the skies over Billings, Montana. Following, the Westwind's demise in Santa Ana, the National Transportation Safety Board conducted a full investigation that prompted the Federal Aviation Administration to insist air traffic controllers warn pilots about the high level of turbulence generated by 757s and recommend that smaller planes maintain a distance of five miles when trailing one.

That was of course little consolation to the Snyders, the families of Phil West and Jack Sims, and the thousands of associates at In-N-Out Burger. Flags at all ninety-three stores flew at half-staff. Disbelief seemed to hang over the Southland like a blanket of brown, hazy smog.

Rich and Phil West had known each other since childhood; West's grandmother lived across the street from the Snyders in San Dimas. They grew up together in the halcyon postwar San Gabriel Valley, playing army in the avocado fields, chasing girls, and going to In-N-Out Burger. Thirty-seven-year-old West lived in Glendora near Esther Snyder with his wife, Lori, and their young son. He had worked at Snyder Distributing through high school. At Esther's urging, he had gone to work for In-N-Out corporate during college, eventually rising to the rank of executive vice president of administration. Esther considered West another son, calling him "my best friend in the company." He was Rich's right hand. Uncommonly grounded, West could be counted on to defuse the chronic tension between the Snyder brothers.

Jack Sims, forty-seven, another longtime friend, was heavily involved in helping Rich produce *Burger Television*. Along with Richard Rossi, an actor and ordained minister, in 1986 Sims launched a popular but controversial church called Matthew's Party. A church for people who ordinarily disdained churchgoing, the Party often met in a sports gym or a bar where wine and snacks were served, rock music was played, and anyone in need was given money.

Sims and West regularly attended In-N-Out's annual Montana fishing trips, while Rich and Sims met weekly for Bible study. In fact, Rich considered West such a close friend that five days before the crash, on December 10, he had a trust drawn up earmarking a portion of his estate for him and a small circle of other business associates in the event of his death.

Widowed after barely one and half years of marriage, Christina Snyder was shattered. "To me, it felt like I lost an entire volume of life," she recalled plaintively. "If just Rich had died, I would have gone to those two men for support and memories. The loss was just incredible." Just two days after the crash, Rick Plate showed up at Christina's house with an eight-week-old golden retriever wearing a red bow. The dog was to have been Rich's surprise Christmas gift to his wife. She named the puppy Harry Snyder.

In the bewildering days immediately following the accident, In-N-Out Burger said little publicly, preferring to remain, as usual, incredibly tight-lipped. They did release a statement that read: "[Richard Snyder] was the type of person who did a lot more than just talk about taking care of his associates. He nurtured them and wanted them to know how very important they are to the success of the company. Thanks to Richard Snyder, In-N-Out Burger is a great place to work. . . . Richard Snyder has given countless numbers of people the opportunity to lead better lives."

Stories of Rich's quiet and compassionate generosity began to circulate. "He was so obviously a good man," said Rich's friend Bruce Herschensohn, who decades later continued to reel at the random cruelty of the tragedy. "He was just marvelous, through and through." Like many others, Herschensohn could recall numerous acts of kindness that Rich had displayed. "I used to wear these clip-on ties that he just hated. He bought me twenty-five ties and had his secretary knot them. All I had to do was just put one over my head." Most of Rich's good deeds were performed anonymously; every month, Rich sent an In-N-Out cookout trailer to feed the homeless living on Los Angeles's Skid Row.* According to Christina, when her husband found out that an associate at In-N-Out and his wife could not conceive a child, Rich paid for the entire expenses for the couple to adopt—he did the same for another friend.

For days, newspapers and television broadcasts parsed details of

---

* In-N-Out continued this practice following Rich Snyder's death.

the tragic crash and spent a great deal of time speculating as to the fate of the privately held, family-owned chain. Rich's death came at a crucial time in In-N-Out Burger's forty-five-year history. He had transformed the small local burger joint into a growing empire that went head-to-head with the corporate giants in the fast-food industry. In fact, secretly, In-N-Out was the envy of the industry. The ninety-three-store chain was pulling in about $116 million annually and employed roughly three thousand associates.

In the final stages of relocating its corporate headquarters to Orange County, the company decided to scale back the scope of the transfer and sent only about fifty executives and administrative employees to Irvine, keeping most of the corporate associates in Baldwin Park. Without warning, In-N-Out Burger now faced a host of challenges starting with its succession, transition, leadership—its future.

But those questions were briefly put on hold. In an uncharacteristic display, a public memorial service was held for Rich Snyder, Phil West, and Jack Sims at the Calvary Chapel in Costa Mesa. The service was conducted on Thursday morning, December 23. Nearly three thousand people crowded into the church's pews. In addition to family and friends, the mourners included countless In-N-Out Burger associates as well as business and community leaders, elected officials, and members of the Orange County Police Department, the California Highway Patrol, and police officers from nearly every community that In-N-Out Burger had touched. "There were more owners of restaurant chains then you can imagine," recalled Jack Williams. "It was like a restaurant owner's convention. They all wanted to pay tribute to Rich." In fact Williams and his wife, Linda, did just that two years earlier when they opened up a new restaurant near their ranch. They called it Richie's All-American Diner. Irvine mayor Sally Anne Sheridan arrived at the church early only to find that all of the seats were taken, and she ended up standing through the entire service. "It was incredible," remembered Sheridan, "very religious and moving."

Wreaths of flowers filled the church's lobby. One was fashioned out of red and yellow carnations forming the In-N-Out logo; another was in the shape of a cheeseburger. Enlarged photographs of the three men graced the pulpit on each side of the speaker's podium. Numerous condolences were sent, including those from former presidents Ronald Reagan and Richard Nixon and California governor Pete Wilson. An In-N-Out cookout trailer was there to serve burgers to the guests following the service. Gracing the cover of the memorial service's program was a photograph taken of the three men only eighteen months earlier at Rich's wedding.

Several years earlier, Rich was so upset after attending the funeral of a dear friend that he felt did not represent her in the least that he immediately outlined exactly the kind of memorial he wanted, down to the music to be played and Bible passages to be read; his memorial followed his instructions exactly. The service was punctuated with Scripture readings and tales of the three boyhood friends. A recording of country crooner Lee Greenwood's anthem "I'm Proud to Be an American" was played. Longtime store manager Don Miller, who had come to regard Rich as a brother, described how he made Rich "cry like a baby" teaching him to slice onions while training him some twenty-five years earlier. "That man," he said, holding back tears of his own, "was a legend in my mind."

Chuck Smith Jr., the pastor who had presided over the recent wedding of Rich and Christina, told the assembled crowd, "To miss them is certainly understandable. But to remember them without a smile would be a crime."

When Esther rose to speak, she was met with two standing ovations. Her face creased with sadness and her voice at times breaking, she described her last moments with her son before she left him at Brackett Field. "Richard said, 'Mom, I'm so glad you got to go with us today.' I kissed and hugged him. When I got off the plane, Phil was there with my jacket. I sat back and waved and thought, 'Lord, I'm glad I got to know these men.'" Joe McCaron recalled Esther's stoic display. "I never saw her get emotional," he said. "Even at the funeral she kept herself together, and that's pretty hard to do." Despite her obvious pain, Esther did not want anyone to feel sorry for her, and she focused on how her daughter-in-law was coping.

Rich's widow, Christina Snyder, took to the podium along with the widows of Phil West and Jack Sims. Looking out into the crowd with incredible composure she said, "Right now, as our hearts are grieving, and we feel empty inside, the only hope is to know that Jesus is with us. He's our strength right now."

Guy Snyder's relationship with his younger brother had been complicated. By turns they were at odds and often on the outs. The two men lived almost entirely different lives. The brothers did, however, share a history, and that history included a strong pride in In-N-Out and what their parents had built.

At the time of Rich's death, the two brothers were barely on speaking terms—but when Guy received the news, he was absolutely stricken. He dissolved into tears. "I'll never forget the night he called me and told me that Rich had died," recalled a close friend. "He was just hysterical. He went down to Newport Beach and set things up in a hotel and took over. He stayed there until the funeral."

Facing the packed church, Guy shrugged off his shyness. In a rambling eulogy, he alluded to the accident and spoke of "God's work." Touching on their pronounced troubles, Guy confessed, "I thought I might be bitter." Then, focusing on the positive, Guy praised his younger brother. "Richie and I sometimes couldn't even sit down and talk about things," he said. "Richie always stuck up for me. I always did everything first. . . . This one time, Richie went first. What happened has changed a lot of people's lives." He added, "If I can follow in his footsteps, I will be the happiest person in the world."

Under Rich, In-N-Out had solidified its position as a player in the industry, and it enjoyed an enviable cult status among its customers. It had come to personify the Los Angeles lifestyle with its humble blend of fast food and car culture. Rich imbued the family firm with his effusive personality and nimbly transformed the restaurants into a big business without damaging the integrity of the small burger chain that his parents had built. Seemingly gifted with both the Midas touch and the common touch, under his guidance, In-N-Out was in good hands. He was that rare family scion who was not only capable of running the business but was able to carry it to greater heights. His death left not a power vacuum but was a real blow. As Esther later confided, "When Richard was killed . . . my world had ended, almost. I had never had to worry about anything as long as he was here. He was a happy soul."

A ripple of uneasiness spread within the company. The sudden loss of such a charismatic and forceful leader left longtime associates concerned about what would happen next. At seventy-three years old, Esther Snyder did not appear a likely candidate to run In-N-Out despite her enthusiasm, energy, and love for the company that she had helped to build. She still worked in accounting five days a week and attended numerous company events and store openings, but at her advanced age she did not appear to be the first choice to take over.

And the chain's second choice, Guy Snyder, had a big question mark hovering over him. During the past seventeen years, as his younger brother ran In-N-Out Burger, Guy had been kept at arm's length. The tension between the Snyder brothers was legendary, and in the years of major expansion and transition at In-N-Out, Guy was largely out of the picture. As the years passed, he had less and less to do with the operations of In-N-Out Burger. His efforts notwithstanding, Guy still suffered from his crippling addictions and bouts of depression. Without Rich in charge, many questioned whether the family would go ahead and sell off In-N-Out Burger.

In the days and weeks following the memorial service, company officials indicated only that out of respect for the family, any decision about succession would be "delayed indefinitely." For the moment, the company remained in a state of mourning. The considerable management vacuum created by the catastrophic deaths of Rich Snyder and Phil West left a gaping hole. Since Baldwin Park maintained its silence, numerous outsiders stepped in, all too happy to speculate on what would happen next.

A gaggle of business consultants and others with little to no inside knowledge offered their insights on whether Rich had a succession plan in place or not, and what that might mean for the future. The most popular theory was that In-N-Out Burger would be sold.

In those days following Rich's death, the phone at corporate headquarters in Baldwin Park rang incessantly with calls from prospective buyers who wanted nothing more than to pounce on what they perceived to be the vulnerable carcass of the Snyder family. A number of interested parties approached In-N-Out Burger, testing the waters. Reportedly, one of them was PepsiCo, which had for years been snapping up fast-food chains across the country.

A cacophony of thinly informed speculation emanating from analysts, bankers, and lawyers proclaimed that an announcement was imminent. Whispers turned to loud rumors; In-N-Out was on the block. In truth, any company would have been happy to get its hands on In-N-Out. Well-run and flush with profits, In-N-Out had not only managed to grow and develop with its unblemished reputation intact; In-N-Out was an unassailable brand name.

In reality, In-N-Out Burger was never on the market. There was never any suggestion that it would be sold—at least not inside the Baldwin Park headquarters. For all of the talk among industry watchers, the one thing they had no understanding of was the will of Esther Snyder. Those closest to her said that her devotion to the company and her loyalty to its associates trumped any possible consideration of a sale. She was primarily concerned with what would happen to them if In-N-Out were sold or made part of a larger company. Besides, Esther knew that Rich would never have sold the company. Despite the incessant murmurings on the outside, inside the lavishly appointed executive offices on the second floor of the Baldwin Park headquarters, the company never entertained any kind of serious offer.

In early January 1994, In-N-Out Burger had come to a decision. At seventy-three, Esther Snyder stepped in as president, assuming a larger role in the chain's day-to-day operations; Guy Snyder returned to the company as well. He was named chairman, a newly created title. Christina Snyder, Rich's widow, joined the firm's board of directors. Corporate conglomerates and investment bankers had once again been given the cold shoulder. The family chain stayed in family hands. In the minds of In-N-Out's legion of fans, their Double-Doubles were safe from external meddling.

Inside In-N-Out Burger's Baldwin Park headquarters, the news had a slightly different flavor. Among immediate family, close associates, and the top management, Guy Snyder's appointment was viewed with something less than exuberance or even relief and more along the lines of cautious trepidation. Quite simply, his return to Baldwin Park was not seen as that of a wunderkind in the mold of Rich Snyder, but rather as the triumph of the prodigal son after long years spent in the shadow of his younger brother.

Rich had been actively negotiating over the course of the previous year to buy out his brother's outstanding In-N-Out shares in order to assume full control of the family business. He had strongly urged Guy to retire from the company on his own. "He wanted him out of his hair," was how one observer put it. Rich spent the year before he died pursuing a legal framework that would keep Guy permanently out of the company—but despite his own recurrent troubles, Guy was not one to be shunted aside so easily. As for Esther, she just wanted peace between her sons.

During this time, Andrew Puzder was Esther's attorney. The role put him in a position where he observed, as did many others, the complicated rhythms of the Snyder brothers' relationship. A former trial lawyer, he recalled that in the year before the crash, Esther and Rich "were trying to work out an arrangement between Guy and Rich about how to go forward with the business." Esther still owned the majority of the company, but Rich was looking to formalize his role in running In-N-Out Burger in a way that would leave Guy out. "I represented Esther's interests and was trying to help her," explained Puzder. "She didn't want to alienate either son, and wanted the process to go smoothly."

The situation added another layer of strain to an already tense relationship. Although Guy had been on the outside, he still had an emotional as well as financial interest in the burger chain. Since childhood, the brothers had been fiercely competitive with each other. Whatever the circumstances, neither of them wanted to be shown up. "Both of them wanted to be treated fairly and honorably," explained Puzder. "They were both concerned with protecting the company and protecting their interests. Both wanted to walk away without losing face"—although it seemed that the latter was more of a concern for Guy than his brother. As Puzder noted, In-N-Out Burger "had really become Rich's brand."

CHAPTER 18

As it turned out, on December 14, Rich had traveled up north along with his wife and Esther to see Lynsi perform in her school's Christmas pageant. Rich was crazy about his eleven-year-old niece. Every year since she was two years old, he had made a special date for the two of them to spend a day at Disneyland. While in Shingletown, he had hoped to talk to his brother. It was an awkward visit that also happened to fall on the anniversary of Harry's death. The relationship between the two brothers, already filled with resentment and jealousy, was at the breaking point over the buyout. The legal documents to buy out Guy's shares of In-N-Out and remove him as a trustee from the family trusts were in the process of being finalized.

Rich already had a much larger stake in the company than his brother, and in three years, the majority of In-N-Out shares would be transferred to Rich; Guy had little choice but to sign on the dotted line. Although the deal ensured his financial security for life, the sting of the situation left him feeling like the victim of a great injustice.

The year-long buyout proceedings had left a bitter taste in Snyder mouths, and the aftermath promised more of the same. Family relations were already badly damaged, and the fear was that once Guy signed, those ties might be irrevocably severed. For that very reason, Esther had asked that the documents not be signed until after Christmas.

During Lynsi's pageant, Rich pulled Guy aside. "We might not see each other again," he told him. "You're my brother, and I love you." Rich, Christina, and Esther Snyder flew back to Southern California. Less than twenty-four hours later, Rich was dead.

Without a legal arrangement in place removing Guy Snyder or naming an alternate successor, the terms of the irrevocable Esther L. Snyder Trust remained in effect—and the terms were unambiguous. In the event that Rich died without any living descendants, the majority shares of the trust—in effect the majority of In-N-Out—were to be transferred to Guy Snyder. Rich and Christina did not have any children of their own, and Rich had not adopted his wife's daughter, Siobhan.

Among the many tragedies and pieces of unfinished business that resulted from the crash was the fact that Phil West was onboard the Westwind that day; that was the crucial event that turned control of the company over to Guy. The matrix of trusts and estate planning instruments created to protect In-N-Out and ensure its succession through Rich unraveled. West was named as the successor trustee of the Esther L. Snyder Trust after Rich, as well as a cotrustee of Rich's own trust. Had he not died on the plane, West would have administered the majority of the company shares that comprised the core of that trust, set to roll over in three years. Regardless of the buyout, the move would have prevented Guy from gaining control over the company.

With West gone, Guy's place as trustee was secured. By an act of cruel fate, Guy, who had only recently been on the outside looking in, was now the man holding the keys to the kingdom.

Legally, Guy Snyder was in control. On the surface, In-N-Out seemed destined to implode. Guy's newly minted leadership remained a great unknown. Even apart from his long catalog of turbulent personal problems, Guy was viewed as something of an uncertain successor. In part, this was because Rich had proved such an exceptional leader; a vibrant, charismatic president, Rich had led In-N-Out Burger to great success largely on the basis of his personality. He had quintupled the size of the chain, increasing profits handily. He managed do so without diluting the unique character that had turned the burger chain into a cult phenomenon. "When Rich took it on, it was a nice little place with a '50s style," was how Andrew Puzder described it. "He really marketed it and played it up." During her younger son's time at the top, Esther had played a largely supporting role, leaving the final operational decisions to Rich.

While Guy stood at the company's edge, he had done little to assert himself or insinuate himself into a greater role. Those who knew Guy Snyder uniformly commented on the fact that he always seemed happiest on his ranch or at the racetrack. He loved nothing more than to put on a pair of overalls and work the land in one of his tractors or old farm equipment. His ultimate dream was to live up on the five-thousand-acre company-owned ranch in Arroyo Grande near San Luis Obispo and raise cattle.

Despite the anger and resentment that pricked under his skin, when Guy was named chairman, it was as much of a shock to him as to anyone else. "He'd been pushed out of everything, pushed out in Shingletown," was how one of his friends described the situation. "He was a little pissed off, but after a while he wound up on the ranch with Lynda and Lynsi, and he had adjusted to that lifestyle." Although he hadn't signed away his rights to the company, Guy had been on the periphery for so long that he wasn't too sure he wanted it, and certainly not under these circumstances. "When Guy came back, the company was drop-kicked into his lap," said the friend. As Guy told the mourners at his brother's memorial, "My life has changed quite a bit. I never thought things would go down like this."

In-N-Out's executive team had to view Guy Snyder's installation as company chairman warily. Few had many interactions with him while Rich was in charge. They seemed "skeptical and a little fearful at the prospect," said the friend. "Some of the management had been talking before Guy actually took over." The conventional wisdom seemed to be "that Guy would probably get tired of running the company and would go back to his ranch in Northern California and let the others run the company."

It may have been a case of wishful thinking, at least in the beginning. Guy's chairmanship dovetailed with a relatively healthy stretch in his life; he seemed to have his chronic troubles under control. Beneath the fog of painkillers, many remarked that, actually, Guy was an incredible fellow. Lucid, he was a solid manager who displayed sound judgment. In particular, he was said to have a phenomenal sense about people. "He could read them really well," said one. And he understood the product better than anyone. Ironically, in spite of his own issues, Guy seemed to be able to handle external problems fairly well. Moreover, he wasn't cavalier when it came to the family business. He had a real understanding of the responsibility now placed on his shoulders. When Guy came on board, proclaimed the insider, "He didn't make any blunders and was actually quite knowledgeable."

One of the first acts of In-N-Out's board was to raise Guy's annual salary to approximately $2 million a year. One of the first things that Guy did was to run out and purchase some suits for his new job. It was a half-hearted gesture, however, and he quickly lapsed back into wearing his customary jeans and T-shirts. His casual attire notwithstanding, Guy seemed to be fairly clear-eyed in his view of In-N-Out Burger and its future. Still, he had lot to learn about the chain's operations and management.

Rich had plotted out a growth plan that would continue to extend the chain's reach. In principle, under his strategy, In-N-Out could have continued moving (slowly) toward becoming a national chain (the real question was whether that was his intention). Rich was a realist, and given the chain's delivery of fresh product and management pipeline, he never thought In-N-Out Burger would move beyond the western United States during his lifetime.* The plan under way was to roll out ten to twelve new stores a year regionally while following the chain's longtime strict quality guidelines and then review the situation before expanding further.

When Guy stepped in, the company was on the cusp of implementing one of Rich's five-year growth schemes and In-N-Out was

---

* One thought Rich did express, however, was that he would not consider putting an In-N-Out in Hawaii. One of his favorite holiday spots, he didn't want to spoil his vacation with work.

on a building binge, having recently established a foothold in Nevada and setting its sights on Arizona. During his first few months on the job, Guy went around visiting most of the ninety-three stores, checking up on their operations and meeting with the associates and managers.

Following his tour, Guy came to the conclusion that the five-year plan was too aggressive, and he wanted to scale back the planned roll-out. In-N-Out Burger grew only as fast as its management strength would allow, and Guy didn't see enough quality people who were ready become store managers. At the current pace, he believed that the managers would be stretched too thin. One of his first executive decisions was to slow down the chain's expansion, capping new openings at eight or nine stores a year. In a rare public comment, Esther later validated her son's judgment. "Guy believes we need to have the time to train people properly to run the new restaurants," she had told a *Los Angeles Times* reporter, "and he's probably right."

In those first few weeks and months following Rich's death, Guy did not exactly tread lightly. He moved into his brother's corner office, the largest of the executive suites in Baldwin Park. The company didn't abandon its planned transfer to Irvine, and a number of its executives moved to the Irvine University Tower. Guy had Rich's Irvine office redecorated, but he only worked there when it was absolutely necessary; he much preferred to work out of the original headquarters. The commute from Glendora was a brutal one, and the Irvine headquarters represented his brother Rich. Guy felt more at home in Baldwin Park where he had grown up. The warehouse, processing, and distribution center remained in the original location—and that was exactly where Guy's real strengths lay. He seemed to have a kind of sixth sense when it came to the smallest details of the warehouse operations and the product. Some marveled at how Guy could look at the sponge dough or the secret sauce or the milkshake mix and immediately tell that there was something off about the product even when most others could not.

Whether he was working out of the original Baldwin Park head-quarters or down in Irvine, Guy was surrounded by executive managers and consultants who had been put in place by his brother. Perhaps it came as no surprise that he wanted to clean house to some degree. One of the first to go was Andrew Puzder. "As soon as Rich died I was out of the inner circle," he explained. "I represented Esther and sided with Rich." After Guy took over as chairman of In-N-Out Burger, he said, "I was kind of out of it." In something of an ironic twist, Puzder went on to work for Carl's Jr. founder Carl Karcher. After representing Karcher in an insider trading case (later settled with the SEC), Puzder became general counsel for Carl Karcher Enterprises. And in 2000, he was named the president and chairman of the company.

Apparently, Guy was singularly unhappy with the decision to put his brother's widow, Christina, on In-N-Out's board of directors. The two were said to have a rocky relationship. Guy had indicated to some that he didn't regard her as really part of the Snyder family—after all, she had been married to Rich for less than two years. He began agitating for her removal. Esther had no problem with Christina as a board member, and the two remained close. Emotionally sustaining one another during the difficult first year after Rich's death, Esther often stayed at Christina's house overnight for comfort. However, exhausted, non-confrontational, and perhaps even a bit guilty, she did not oppose her son.

Within two years, arrangements were made and an undisclosed settlement was agreed upon. Christina Snyder resigned from the board. It was a situation that she later described as "painful and draining." Christina headed up In-N-Out's Child Abuse Fund, the charitable foundation that Esther and Rich Snyder had founded in 1984. (In 1997, she renamed it the Rich Snyder Foundation.) As a result, the chain ended up establishing a second corporate charity called the In-N-Out Burger Foundation.

After Christina Snyder left, Guy turned his attention to a group of the chain's executive vice presidents whom he wanted gone. It was a thorny problem. Before Rich died, he set up a trust that gave a small percentage of In-N-Out shares to a handful of executives (including Phil West) in the event of his death. These particular shares came from the stock that Rich had inherited from Harry. Prophetically, the decision stemmed from Rich's understanding that if something were to happen to him, he could ensure a consistent and smooth transition if his talented executive team stayed in place. The shares were offered as an incentive to keep them and set up to give the vice presidents a percentage if they remained for a certain period of time. Rich also arranged to give a few other close friends who were also longtime associates a small portion of shares. In the latter case, the shares were more of a reward for loyalty than motivation to stay put.

In a move that severely angered Guy, in the summer of 1996, four of the company's executives filed suit against the Los Angeles law firm that had crafted the trust. They alleged breach of contract and malpractice and sought $1.5 million from the firm in damages. According to the *Los Angeles Times*, the four men claimed that "the law firm's trust work resulted in 'significant confusion, ambiguity and expense' that has caused them 'significant' but unspecified monetary damages."

The lawsuit kicked Guy's ire into high gear. Not only did it air In-N-Out's private business publicly, Guy didn't feel that some of the executives were entitled to a piece of the company to begin with. Unlike Rich, Guy had a personal problem with this group of college-educated, professional managers who had joined the company only a handful of years earlier. In his mind, they hadn't helped to build up In-N-Out over the decades, and if anybody deserved a stake in the company, it was family or those that had started at the bottom picking up trash, working construction, or performing maintenance and worked their way up through the years.

In a fit of pique, Guy wanted to get rid of them. In the end, however, he was prevailed upon not to clean house. He was convinced that clearing the executive decks of such capable and dedicated managers would not be in the best interest of the company. In fact, it was likely to have a devastating effect. Denuding the chain's organizational chart of its highest-ranking members would have been a shock to the smooth-running operations and would no doubt have jolted loyal associates down the ladder who had been with the company for years. In the end, the executives stayed put, as did the tension between Guy and In-N-Out's top managers.

Perhaps to everyone's great relief, Guy did not seek to remake In-N-Out Burger. For the most part, he was content to keep things running just as they had been. As a result of the infrastructure that Rich Snyder had put into place, the leadership transition—at least in terms of operations—was practically seamless. The company could still grow and expand and increase its sales volume by following the chain's philosophy of "Quality, Cleanliness, and Service" and the systems put in place to achieve it. During his tenure, Rich created human resources departments, the university, and the accounting departments, and he enlarged and modernized the warehouse and commissary as well as the departments that supported their operations. He created a management system that could ensure quality even as it added new stores each year. It was not lost on close observers that In-N-Out continued to prosper without its most dynamic leader. All Guy had to do was follow the system already in place.

Rich's own charismatic presence had created a very specific corporate culture. It seemed that his legacy was not the achievements of growth and profits but the simple fact that the company he built continued to succeed without him. Moreover, it was a company where people were happy to come to work every day.

CHAPTER 19

In 1994, Guy Snyder celebrated the opening of In-N-Out's hundredth store in Gilroy, California. A town best known for its mushroom farms and its annual garlic festival, Gilroy was located in the Santa Clara Valley; by the time that In-N-Out unveiled its drive-through there, the chain was generating an estimated $133 million in sales. Store number one hundred had been an important cornerstone of Rich Snyder's expansion plans. But for Guy, the milestone was—much like his chairmanship of In-N-Out Burger—inherited from his late brother Rich.

Under different circumstances, the opening would have been an exuberant affair. Rich was always one to go all out when it came to big events and celebrations of any stripe. But given the circumstances, aside from a cake and festive decorations, the grand opening of the hundredth In-N-Out Burger was a rather subdued occasion.

Guy arrived at the store on 641 Leavesley Road off of the 101 Freeway in the custom-built In-N-Out bus. Something akin to a rock band's touring bus, the plush set of wheels was wired with telephone and computer hookups. It was one of Rich's initiatives. With the chain's expansion continuing full force, Rich thought it was a good idea to have a kind of rolling office that could allow his team to work as they traveled between stores.

Wearing a dark suit and tie and tinted sunglasses, Guy joined his mother, Esther, in the balloon-filled store. At seventy-four, she suffered from chronically bad knees and a heart condition, but Esther still loved to attend as many new store openings as she could. At the Gilroy opening, Esther and Guy cut a red ribbon that proclaimed "100th Store" with a pair of oversized prop scissors.

In-N-Out Burger entered the 1990s as if operating within an alternative fast-food universe. At least in terms of scale, the chain remained a small player in the vast $74.3 billion fast-food industrial complex dominated by a cabal of billion-dollar corporations. As Guy and Esther marked In-N-Out's triumphant march toward Northern California, McDonald's had swelled to thirteen thousand stores with a global push. As the Snyders cut the red ribbon in Gilroy, McDonald's opened new stores in Kuwait and Egypt. Four years later, Burger King, which had opened its first international franchise in the Bahamas in 1966, marked the opening of store number ten thousand in Sydney, Australia.

Harry and Esther Snyder's old friend Carl Karcher had also fueled his chain's growth through expansion. He had taken his Southern California chain public in 1981, and in 1984 opened its first franchise. Six years later, of the 561 Carl's Jr. restaurants, 133 of them were franchises, and the chain was earning $575 million annually. Having opened stores across the West, Karcher began his own global march, opening stores in Mexico and the Pacific Rim. "One of my favorite memories was going international," he proudly exclaimed.

But Karcher took his company in another direction as well during the 1990s. He invested heavily in real estate in Anaheim and the Inland Empire, using his Carl Karcher Enterprises stock as collateral for his bank loans. It was a disastrous gamble. The investments left Karcher with $70 million of debt at the same time that the company's stock was falling. His attempt to bolster sales by acquiring the Mexican restaurant chain the Green Burrito in 1993 was rejected by the company's board, and the board in turn ejected Karcher from the company. Two months later, Karcher wangled his way back in, orchestrating a takeover through a series of intricate financial maneuvers. The turnaround was successful and in 1997, CKE acquired

Hardee's for $327 million, giving the onetime hot dog cart a national footprint. But in the process, Karcher nearly lost everything.

All the same, fast-food consumers had grown fickle. The industry found itself pulling new tricks out of an old hat on a constant basis. After Taco Bell introduced the concept of value pricing in 1988, every big chain followed suit. Each year saw some kind of price slashing battle. In 1996, Taco Bell launched its $1.99 Extreme Value Combos—a year later, McDonald's announced the fifty-five-cent Big Mac (when purchased with a drink and fries), sparking an all-out fast-food price war. As the leading chains cut prices, they also began increasing portion sizes, ushering in the era of super-sizing.

By 1997, the fast-food industry reached $109.5 billion in revenues, closing in on the $114.3 billion in sales generated by sit-down restaurants. The business pages of newspapers across the country were filled with stories about every move the colossal and highly influential industry was making. "McDonald's Still Finds There's Still Plenty of Room to Grow," the *New York Times* reported on January 9, 1994. "Bigger Portions Being Thrown as Global Fast-Food Fight Heats Up," proclaimed the *Press-Telegram* of Long Beach on March 8, 1996.

The chains sought myriad ways of capturing a capricious market. One of the most ambitious efforts was also one of the most disastrous; in an attempt to cater to the adult market, in 1996 McDonald's introduced the Arch Deluxe, a line of burgers and sandwiches with more sophisticated ingredients. The launch was accompanied with a staggering $100 million advertising campaign created by Peter McElligot and with the tagline "The Burger with the Grown-Up Taste." The program was a grown-up debacle. The Arch Deluxe never caught on, and McDonald's soon abandoned the costly enterprise.

Subtle changes were taking place at In-N-Out Burger as well. Around 1995, a raucous discussion broke out over the decision to add a fourth beverage size at the stores. The managers seemed to go back and forth over whether to move forward or not before they ultimately greenlighted the addition. The following year, the chain began serving Dr

Pepper. And In-N-Out opened a new warehouse in Tracy, a kind of satellite distribution center for its fresh ingredients that allowed the chain to serve the growing number of drive-throughs in Northern California. By the end of 1996, there were 116 In-N-Out Burger drive-throughs generating about $159 million in sales.

Back in Baldwin Park, Guy was looking to make some further changes of his own. The friction between him and a handful of top managers had yet to dissipate. Guy seemed to lash out and then cool down—in the heat of the moment he often wanted to cut several team members loose before deciding on a different course of action. During this time, the chain's CFO Steve Tanner* departed, later going on to hold similar positions at several restaurant chains including El Torito.

Like Harry, Guy was a zealot when it came to the chain's "Quality, Cleanliness, and Service" scheme. Moreover, he expected In-N-Out's associates and managers to aggressively handle workplace issues. In one episode when a longtime manager did not (in Guy's estimation) step up to the plate, Guy dressed the man down: "You don't have any balls," he said.

Guy was said to be particularly unhappy with one of the chain's regional managers who had many decades of service to In-N-Out under his belt. Initially, Guy wanted to remove him, but rather than let him go and cause uneasiness within the ranks, Guy was convinced to appoint Mark Taylor (his stepson-in-law) as general manager to supervise the chain's two regional managers. In the organizational structure of In-N-Out Burger, regional managers were in charge of the divisional managers, who in turn supervised seven to eight stores each. It seemed a reasonable solution. Taylor was part of the family, so to speak, and his appointment gave Guy a way to manage the regional managers without making any drastic changes.

Taylor began working at In-N-Out Burger in 1984 when he was nineteen years old. A year later he married Traci Perkins, Guy's stepdaughter. The pair met when they were both students at Bonita High

---

* Tanner served as In-N-Out's CFO from 1991 to 1996.

School (Guy and Rich Snyder's alma mater). Rich found him to be hardworking, and Taylor moved up through the ranks, becoming the store manager of In-N-Out units in Pomona and West Covina.

Guy wasn't sold on the idea at first. He was said to be uncertain about Taylor's skills. While he eventually signed on, he did so with some reservations. As Taylor stepped up to corporate management, it wasn't long before Guy began joking, "He wants my job."

Despite his numerous efforts to stay clean, Guy's troubles soon re-surfaced. It was a topic that the Snyder family and In-N-Out Burger preferred not be examined publicly. Friends insisted that Guy's prob-lems with drugs stemmed from the painkillers he continued to use in the constant search for relief from his back and arm injuries. After a while, however, the painkillers had been supplemented with a cor-nucopia of other substances.

By the mid-1990s, the impact of Guy's problems had become dif-ficult to ignore. Guy had good periods and bad periods. He would get to a point where he could barely function and then enter rehab, but his numerous stays did little to curb his addictions. "The way he looked at rehab," explained one of his friends, was "do those thir-ty days and get out. The whole time he was in there he was craving drugs, and when he'd get out, he couldn't wait to start all over again." At one point, Guy was refused admittance to the Betty Ford Cen-ter because he was too high to talk to the admitting officer. It was a volatile situation that no doubt contributed to what close observers described as a turbulent marriage. Guy was often absent, racing, in rehab, or attending to some other interest of his, and the couple were said to argue frequently. In August 1995, Guy and Lynda Snyder le-gally separated. Their estrangement lasted two years.

Barely four months after the couple's formal separation, police officers in Claremont, thirty miles east of Los Angeles, found Guy Snyder's parked 1968 Dodge Charger. It was around 3:30 a.m. on Christmas Day. When the officers peered into the car, they found Guy hunched over and passed out; his face was buried in a briefcase containing a small pharmacy of drugs, including marijuana, Valium, the sedative Klonopin, and the painkiller codeine. In addition, Guy was carrying a loaded 9 mm Glock semiautomatic handgun, an 8-inch switchblade, and $27,475 in cash. He was arrested and charged with several misdemeanors including public intoxication, possession of less than one ounce of marijuana, carrying a loaded firearm, and possession of a switchblade. Released on his own recognizance, six months later Guy pleaded "no contest" to the weapons charges at the Municipal Courthouse of Pomona. Los Angeles County prosecutors dismissed the drug charges and in turn Snyder paid a fine of $810. As part of his probation, Guy agreed not to own, use, or possess any dangerous or deadly weapons.

The incident did not go over too well back in Baldwin Park. For obvious reasons, news that the chairman of the beloved and wholesome In-N-Out Burger had been found unconscious by the side of the road and arrested for drug possession was not exactly a public relations windfall. Bracing for the worst, In-N-Out's executives convened a meeting about how to handle what they expected would be a deluge of media inquiries. However, the incident remained largely out of the press; not even the chain's local paper, the *San Gabriel Valley Tribune*, ran the story. The inner circle at In-N-Out seemed greatly surprised when the entire episode seemed to pass without comment. In fact, the story of Guy's arrest did not surface publicly until nearly three years later.

Guy was a loner surrounded by enablers; they helped to feed his addiction and then cleaned up his messes. His old racing injuries provided a pretext for the drugs. Few if any seemed willing to stand up to Guy and say no. He had a complicated web of relationships wherein his friends and business associates were one and the same, and nearly all were on his payroll in some fashion. They enjoyed his childlike generosity. Swept up in the kind of life that unlimited funds provided, it might have been easier to turn a blind eye than to say no.

Although clearly troubled and frustrated by her son's problems, Esther tried to remain hopeful that Guy would soon get well. After the Claremont episode, she was said to have been grateful that he wasn't injured or killed. But Esther wasn't a woman known for putting her emotions and feelings on the table. Rather, she held quietly to her faith. Besides, not many were willing to confront Esther and risk upsetting In-N-Out's kindly grandmother. This was a woman who never uttered an unkind word about anybody, preferring to see only the good in people. When Esther did become distressed about something, she would say "oh, goodness," get worked up, and then declare, "I hope the Lord forgives me—I've got my anger up." That was usually the extent of it.

But Guy's drug problem was something that not even Esther could completely ignore. When Guy was in pain, Esther seemed to understand his need for painkillers. Her strategy was to think positive and hope for the best, but his downward spiral was difficult to comprehend. She often called his entourage to make sure that he was okay. More than anything, her approach smacked of unhealthy indulgence. This sweet old lady who knew the cost of a truckload of potatoes seemed determinedly unaware of the price of her son's drug habit.

During Guy's worst periods, he was increasingly absent from In-N-Out Burger. Lynda's nephew Tom Wright had become Guy's right hand man. Among those in the burger chief's retinue, Wright seemed to genuinely care about Guy as a friend. In addition to his responsibilities in asset protection, he had also been named to In-N-Out's board of directors. Wright often ended up acting as a go-between, telegraphing Guy's intentions to the vice presidents during the chairman's absences at corporate meetings. For years, Rick Plate had served a similar function, both supporting and propping up Guy. In 1994, Plate collapsed at In-N-Out's Irvine offices; he had suffered a brain aneurysm, and never regained consciousness.

Guy's absences didn't necessarily interrupt the day-to-day operations. When he was unavailable, the chain's executives usually met and worked out decisions as a group. After Guy returned, he could be prickly and at times paranoid toward the other In-N-Out Burger executives during meetings. Sometimes he seemed to think that some of the managers were acting to supersede him, and he countered by putting them in their place during a meeting. Regardless, it was due in no small part to the talented and loyal executive team that In-N-Out Burger remained on a stable and successful course. It was Guy Snyder's life that was turning volatile, falling into an unpredictable pattern of good days and bad days, bright spots and periods of darkness.

It was during one of Guy's better stretches in 1995, while separated from Lynda, that he reconnected with Kathy Touché (*née* Nissen). Kathy was the girl whom he had had a crush on when he was in high school and she was just thirteen. A divorced mother of three, the petite blonde was working at her father's restaurant, Norm's Hangar, on the tarmac at Brackett Field in La Verne. A local institution, Norm's was known as a great spot to watch planes take off and climb up over the San Gabriel Mountains and for its all-day breakfasts and chatty waitresses. Although many years had passed, the two still shared many interests. As it turned out, they had both experienced profound grief as well. About two weeks after Rich Snyder had died, Kathy's older brother Johnny Nissen was killed while racing his all-terrain vehicle in the California desert.

The pair began talking on the phone and dating in earnest. According to Kathy, Guy would fly down to Brackett Field and meet her at Norm's. This time around, Guy seemed to display none of his high school shyness. On their first date, Guy picked Kathy up in a restored 1936 panel delivery truck for a romantic dinner at El Encanto, a rambling estate tucked into the middle of the Angeles National Forest in nearby Azusa Canyon. In her voice, which sounds a lot like a Lucinda Williams song—molasses over gravel, she explained, "We had a lot in common. We liked street racing and concerts, and we had a lot of fun together." At the time, however, Kathy insisted that she had no idea of the depths of his problem with drugs.

By the time the couple reconnected, Guy had already begun his long, dark descent. But like many addicts, Guy was also something of a magician. According to Kathy, most of the time Guy went into his office and closed the door. For a surprising period of time, he was able to hide his abuse even as it escalated to other substances—in time he was shooting up heroin. Unlike many addicts, Guy also had the means to infinitely enable his use. Surrounded by an entourage of helpmates and aids, Guy had only to snap his fingers to procure a never-ending supply of drugs.

According to some, he was surrounded by sycophants and hangers-on; others say that his closest friends were simply at a loss, unable to do much for Guy except pick up the pieces—or baby-sit a grown man bent on self-destruction, a man to whom a great number of them owed their livelihoods. A small cadre saw to it that he went into rehab and escorted him to hospital emergency rehabs following multiple accidental overdoses. But like all addicts, Guy couldn't hide from the effects of his addiction forever. In January 1996, about a month after his Claremont arrest, Guy had a drug-related heart attack.

During this time, according to friends, his estranged wife, Lynda Snyder, and her children (now grown) took a tough love approach to Guy. They rarely saw him. He was said to be deeply hurt by the rupture; he had helped raise Lynda's daughters, Traci and Terri, and loved them deeply. But nothing seemed to pain him more than the separation from his daughter, Lynsi, who was just starting her teenage years and was living with her mother up in Shingletown. According to intimates, Lynda made a decision to keep Guy away from his daughter because of his drug use. "She was his whole world," said one person close to the family. Dale Wright recalled that the family simply cut their contact with him. "Lynda didn't allow Guy to see Lynsi much. She limited her time. He didn't get to see her very often. I don't know how often, but he wanted to see her more often than he got to see her."

On January 23, 1997, Guy and Lynda Snyder's divorce became final. She retained the Flying Dutchman ranch that (according to the *Orange County Register*) was valued at $1.4 million as well as child and spousal support, the paper reported, of $500,000 a year.

Eight months later, on September 28, Guy married Kathy Touché; he was forty-six and she was forty-two. Just two days before the couple's wedding, Guy gave her a prenuptial agreement to sign. "I told him we should have one," she recalled sharply. "But I said that he better not show up with it the week of the wedding. And that's exactly what he did." As Kathy described it, In-N-Out's lawyers drew up a lengthy agreement that would give her very little in the event of a divorce. "I took it to a lawyer in Newport Beach to read," she said. "He looked up at me and said, 'Does this guy even like you?'" In the end, she signed it. "I wasn't after the Snyders' money, and I thought by signing it I'd be demonstrating my love and loyalty," she explained. "I had a private conversation with Guy, and I said to him, 'You said that you would help to take care of me.' And he knew that I wasn't going to take him for a ride."

Guy and Kathy married in a small ceremony in front of one hundred guests—mostly family and some of the couple's old high school friends—at a church in Newport Beach. Fifteen-year-old Lynsi did not attend the nuptials. The reception was held aboard *The Wild Goose*, a 1942 World War II minesweeper that the actor John Wayne had converted into his private 136-foot luxury yacht. The wedding party cruised off of Lido Isle in Newport Harbor. In a photograph taken at their reception, Kathy, who is wearing a simple white shift and pearls, is seen dancing cheek to cheek with Guy, who is pressed tightly against his new wife, a Champagne flute in each of their hands.

Following the reception, the pair drove off in a classic 1940s Kaiser Steel Willys that Guy had spent five years restoring. He got a one-day road pass from the city to drive the car on the streets of Newport Beach, and the hulking, vintage U.S. military jeep was so loud that it set off all the alarms of the parked cars as they drove away.

After a Hawaiian honeymoon, Guy, Kathy, and her three children moved into a $1 million house in Claremont. An affluent city at the base of the San Gabriel Mountains in the Pomona Valley, Claremont was filled with tree-lined streets, historic buildings, and the group of seven Claremont Colleges. In fact, the city's outsized proportion of trees and residents with advanced degrees had earned it the moniker as the "the city of trees and PhDs." The Snyders' sprawling new house had six bedrooms, nine bathrooms, and a huge yard with a pool, Jacuzzi, and sauna. It was the biggest house that Kathy had ever lived in, and she called it "overwhelming." Actually, she said later, "It was too big. It didn't feel like a home."

During this time, Guy seemed to function well. He was engaged with the business of In-N-Out. One of his goals was to buy back as many of the leased stores as possible. In fact, he converted a section of the Claremont house into an office. When he went to the company's headquarters, Guy commuted by Lear Jet. Often he flew back into Brackett Field and met Kathy, who was still working at Norm's Hangar (which she eventually took over from her father). It wasn't uncommon for Guy to arrive, put on Kathy's apron, and help clean up. "People would've died if they knew the CEO of In-N-Out was bussing tables," she laughed at the memory. "He loved nothing more than to eat a patty melt on the patio and bus tables." At one point, Kathy said, he made an offer to buy Norm's—but she refused. "I said no. You leave my family business alone and I'll leave you yours."

With Lynsi up in Shingletown and Guy in only sporadic contact with his daughter, he focused on life with his new family. Guy bought a junior dragster for Kathy's thirteen-year-old son, Aaron, and flew the boy and Guy's In-N-Out racing team to a track in Phoenix. "That was really neat because their father wasn't too involved in their lives at the time," she recalled. And Guy took his new father-in-law and brother-in-law on Alaskan fishing trips. "They had a ball. He really admired my father," she said. "He was just a big kid. He liked to drive tractors and get stuck in the mud. He'd spend all day getting stuck in the mud. Sometimes I think he'd do that just so he could get pulled out." The family took frequent trips to the cattle ranch in Arroyo Grande. (It was something of a consolation prize, since he had lost the Flying Dutchman in his divorce settlement with Lynda.)

Accompanied by his new wife, Guy Snyder traveled overseas with the managers (and their spouses) who were rewarded each year as part of In-N-Out Burger's 100 Percent Club. Europe—Amsterdam in particular—was a favorite destination. Harry Snyder's family was originally from Holland. In addition to the family connection, Guy was fascinated by the city's cafés, where one could buy marijuana.

One 100 Percent Club trip to Amsterdam was particularly memorable for Guy. Guy and Kathy flew the Concorde to Europe, stopping in New York en route where they stayed at the storied Plaza Hotel's two-level, seven-room presidential suite overlooking Central Park. After New York, the couple flew to London for three days before arriving in Amsterdam for a four-day trip. On the itinerary was a performance of *The Phantom of the Opera* in New York, dinner at a sixteenth-century castle in London, and trips to the tulip farms of Holland.

Prior to this particular trip, Guy had hired a genealogist and a private detective to track down his father's relatives in Amsterdam. Once there, they met for the first time, and the reunited families took a boat cruise and had lunch together. It was during this visit that Guy looked into buying a Sherman tank and a vintage windmill to be sent back home. Once Guy discovered that shipping charges would run in excess of $1 million, however, he changed his mind.

Now that Guy was chairman of In-N-Out, he no longer had his brother looking over his shoulder. One of the first things that he did after Rich died was to step up In-N-Out's involvement in racing. He purchased a suite at the Pomona Fairgrounds, where the Winternationals were held, and liberally passed out tickets to friends and In-N-Out associates.

In 1996, Guy approached Jerry Darien, whom Guy had known since he was kid running elapsed time slips at the Irwindale Raceway in the 1960s. A former teacher, Darien had worked as an announcer at the Dale when Harry Snyder was part owner in the track. The two met at a Mexican restaurant in Pomona. In addition to the Funny Car, Guy was interested in putting an Alcohol Car team together, and he wanted Darien to head up the effort. Guy agreed to sponsor a team. "There was a handshake," recalled Darien, who hadn't lost his sonorous announcer's voice in the intervening years, "and away we went." The team won its first race in Boise, Idaho, a divisional contest, in April 1997. Six months later, at the end of the season, Guy decided to end In-N-Out's association with the Over the Hill Gang.

Two years later, Guy called Darien again and asked him if he could put a Top Fuel team together. Darien was given three months to assemble a car and enlist a driver. Guy agreed to pay somewhere between $1 million and $2 million for the sponsorship, to be distributed on a quarterly basis. Darien tapped twenty-six-year-old Melanie Troxel, who earlier raced the Alcohol Car, to drive the Top Fuel car. She was one of the NHRA's fastest female dragsters (and the daughter of the 1988 NHRA Alcohol Dragster champion Mike Troxel). The team's first race was to be at the Winternationals in Pomona, in February 2000.

CHAPTER 20

Guy's time at the top was short-lived. His troubles soon escalated, taking away from his ability to function—let alone fulfill his duties as the company's chairman. Simply put, the years of abuse had caught up with him. Guy rarely checked on the warehouse or meat departments anymore. He was spending less and less time at the company, and meetings that would once have been postponed to accommodate his absences now went on without him. The professional management team that Rich had put in place, those men whom Esther referred to as "my boys," implemented the company's agenda with her guidance. In the final years of the 1990s, anyone could see that Guy was in bad shape.

On October 27, 1997, Tom Wright was stopped at the San Ysidro Port of entry near San Diego at the Mexican border. There, U.S. customs agents searched Wright's 1997 In-N-Out–owned Ford Explorer, where they found a cache of drugs that included fifty 120-milliliter bottles of a codeine/ephedrine mixture; four bottles of 21-milligram Rohypnol; approximately 1,000 tablets of muscle relaxants; and miscellaneous containers of inhalants and nasal spray.* According to arrest documents, Wright told the agents that he had traveled to Tijuana

---

* In Mexico, most of these drugs are available on an over-the-counter basis.

to purchase "medicine from a nearby pharmacy for his personal use," paying roughly $1,500 for the supply.

Wright's wife, Dale, however, said that her husband had gone down to Tijuana at Guy's request in order to buy a number of sleeping pills, asthma inhalers, and other prescription drugs for his use. "It wasn't the first time," she said. "But it was his last."

Dale, who had accompanied her husband a few times on these trips to Mexico, said that in hindsight "I didn't care for [Tom] doing it, but it was part of his job, being Guy Snyder's assistant. It was something he had to do or take the results and hit the road. We couldn't help it, and it was better than trying to get them all the time through different doctors."

Years later, Dale still shudders at the memory of what she called her husband's first brush with the wrong side of the law. "It was just horrible," she said. "We were afraid that Tom would go to jail." Guy told the couple not to worry; he assured them that he would pay for any of the financial fallout from the episode. For a time, they kept the entire affair under wraps. The Explorer was impounded, and, as Dale recalled, they explained its disappearance to company managers by telling them the engine blew up.

In December, two months after his arrest, Tom Wright went before the U.S. District Court in San Diego. He was charged with illegal importation of a controlled substance, pled guilty to misdemeanor possession of codeine, and was sentenced to one year's unsupervised probation and fined $1,000. According to his wife, Wright had to pay an additional $40,000 to retrieve the Explorer and about another $100,000 went toward legal fees. Guy kept his word, said Dale, and paid for the entire mess.

By then, Guy's continuing decline was hard to ignore. It was around this time that Guy was diagnosed with porphyria, a complex medical condition often triggered by alcohol and drug use and associated with a variety of ailments including nervous disorders, a severe skin condition, and blue urine. The disease, often cited as the source of King George III of England's mental illness, left Guy's hands covered with open sores that needed to be frequently cleaned and wrapped.

It appeared that even Guy could not overlook the fact that he was trapped in a downward spiral. Just four days before Wright's arrest, in one of his more cogent moments, Guy made arrangements to obtain a legal order appointing successor trustees to the Esther L. Snyder Trust in the event that he died, became incapacitated, or failed to serve as trustee before his daughter, Lynsi, turned thirty (when she was to receive the second installment of the trust, giving her a majority ownership of the company). Guy apparently realized that he needed to do something to protect his only child and the only living direct descendant of the Snyder family. A year earlier, the terms of the trust transferred the majority of In-N-Out's shares to Guy Snyder. His daughter was just a teenager, and she was the sole beneficiary of those trusts worth hundreds of millions of dollars; Lynsi alone stood to inherit In-N-Out Burger.

Three cotrustees were named: Douglas K. Ammerman, an accountant with Peat Marwick; Richard Boyd, longtime Snyder family friend, In-N-Out board member, and the company's vice president of real estate and development; and Tom Wright. In the event that Ammerman, Boyd, or Wright should cease or fail to act as a cotrustee, Mark Taylor, Guy's former stepson-in-law and the chain's director of operations, was named to serve with the remaining trustees. On October 23, 1997, Esther Snyder, the trustor, signed the order before Judge George Olafson of Los Angeles Superior Court.

The new agreement was set up in such a way that no new successor trustees could be subsequently named. If any two of the trustees failed or ceased to act, then the remaining two would serve as cotrustees. Likewise, should the three named cease or fail to act as cotrustees, then the remaining cotrustee would become the sole trustee. It was a straightforward order intended to protect the entwined futures of In-N-Out Burger and its young heiress, Lynsi Snyder. It was this seemingly clear-cut order that later emerged as a factor in an ugly and protracted internal squabble that proved to be a major turning point in the private, family-owned company's history.

The point of contention hinged in part on the naming of Mark Taylor as a replacement successor. During this time, among intimates (including his wife, Kathy), Guy questioned Taylor's abilities and seemed to grow uncomfortable with what he believed was Taylor's growing ambition. According to Kathy, "Guy had every intention of firing Mark." Others concluded that it was Lynsi's mother, Lynda, pulling the strings up in Redding, who agitated to install Taylor, her son-in-law, as a successor cotrustee. Whatever the case, Guy was in no position to put up much of a fight. For Guy, time was running out.

In-N-Out provided a stark contrast to Guy's unspooling life. In February 1997, the chain unseated Wendy's Old Fashioned Hamburgers after eight years at the top of *Restaurants & Institutions* magazine's annual survey of the nation's best burger chain. It also happened to be the first year that In-N-Out had qualified for inclusion. It was a feat all the more remarkable because at the time, Wendy's had about 4,757 stores across the United States, and In-N-Out, with 124 stores in just two states, was competing with all of the large national chains, edging out Burger King (third) and McDonald's (sixth).

The same year In-N-Out earned the *Restaurants & Institutions* recognition, Esther Snyder was singularly honored. At seventy-seven, she was inducted into the California Restaurant Association's Hall of Fame. A year later, the U.S. Navy Memorial Foundation bestowed Esther with its Lone Sailor award. Given to those who have distinguished themselves following their navy careers, demonstrating honor, courage, and commitment, the Lone Sailor had been awarded previously to President John F. Kennedy, former *Washington Post* editor Benjamin Bradlee, author Herman Wouk, and Senator John McCain.

The company was moving forward with new plans of its own. After more than four decades spent alongside freeway off-ramps and suburban intersections, In-N-Out Burger's management team decided, as the turn of the millennium approached, to push further into select urban areas. In a radical break with tradition, In-N-Out unveiled its boldly designed store number 119 in Westwood on the edge of the UCLA campus. It looked nothing like any In-N-Out store that had come before. Stephen Kanner, a celebrated Los Angeles architect known for his pop modern style, created a sculptural burst of bright red and yellow with soaring, sweeping boomerang angles. The neon In-N-Out sign appeared to jut through the cantilevered roof and the glass drive-through allowed motorists to see inside the kitchen from their cars. From the street, the whole building served as a kind of billboard. The store, which Kanner called "a dream commission," quickly racked up a slew of architectural awards.

The chain was now looking to position itself into other high-density and heavily traveled locations. A far cry from the outlying areas where it had traditionally planted its crossed palm trees, In-N-Out was increasingly moving, at least as far as its locations were concerned, from the outside in.

Kanner hoped that the new store marked the beginning of a new visual brand for the chain. "I got really excited," he recalled, and he came up with a series of computer prototypes that would allow the chain to adapt the Westwood design to other locations and still be cost competitive. However, it was not to be. Kanner was told that the Westwood restaurant was too visual. Back at In-N-Out headquarters, Kanner lamented, "They said the burgers should be the star."

As Guy's tenuous hold on his professional activities slipped, his marriage to Kathy was quickly unraveling as well. According to Kathy, the main problem was Guy's louche activities. She had grown increasingly disenchanted with the marriage and felt betrayed by her husband's escalating drug use. Fingering the one-of-a-kind, solid gold Double-Double hamburger pendant around her neck that Guy had commissioned for her, she said, "I feel so stupid, I didn't see the signs," clearly anguished. "I had no clue until after we were married—the severity of it. It was like Elvis. Nobody stopped him. Nobody said no to Guy."

On April 28, 1998, Guy Snyder filed for divorce. The two briefly reconciled—"I kept hoping things would get better," Kathy said—but then on May 18, 1999, he once again filed for divorce. She moved out of the large Claremont house they shared and into a smaller house nearby. "It was pretty mutual and amicable," she said. "Guy really had drug issues and I did not want to be subjected to that life he chose." During their brief marriage, Guy seesawed between rehab and overdoses. "You can't make a marriage like that," she said, looking back. "I know he'd have given anything to come to terms with it." Her tone softening she added, "He was wonderful when he was clean and sober. He was fun, and he was *sooo* handsome."

The couple's divorce moved fairly quickly. The prenuptial that Kathy had signed gave her little leverage. All of the couple's property, according to Kathy, was in In-N-Out's name: "our house, the ranch up in Arroyo Grande, and all of our cars. In-N-Out had their hands on everything." While at court in Orange County, Kathy said the judge displayed little sympathy for her. "He told me that I should consider myself lucky that I got to live the lifestyle of a millionaire's wife while I had the time," she laughed. In the end, she said, "I left with more than I went in with and a lot of wonderful memories."

In his last years, one would hardly guess that Guy was the scion of a multimillion-dollar company. Scruffy and unshaven, bloated and overweight, he resembled a vagrant. Guy no longer cut the dashing figure he once had as a young man in denim with a silver belt buckle and a purposeful stride. Disappearing behind sunglasses and under baseball caps, large T-shirts, and baggy shorts, he was lapsing into a state of dysfunction. In the spring of 1999, he came down with pneumonia, and before the year was up he had suffered and recovered from three drug overdoses. As his behavior grew increasingly erratic, Guy was prone to memory lapses and simply wandering off.

Largely estranged from his family, Guy was worn down and felt increasingly isolated. He moved out of the spacious home he had shared with Kathy in Claremont, and the millionaire who could check into any luxury hotel instead holed up in his motor trailer that was parked inside the old In-N-Out Burger warehouse in Baldwin Park.

At times, and briefly following his divorce, Guy would stay up at the Flying Dutchman Ranch, hoping to see his daughter. Ironically, throughout 1999, Lynda was involved in Guy's hospitalizations after his bout with pneumonia as well as his overdoses.

Toward the tail end of the summer of 1999, Guy once again flew up to Shingletown in an attempt to see his daughter, installing himself in a guest house on the property at the Flying Dutchman Ranch where she was living. According to Dale Wright, Lynda was not happy about this and called her husband, Tom, in the middle of the night and asked him to fly up and take Guy away. "She told Tom that she didn't want him dying up there on her, and asked Tom to come and get him," recalled Dale. Wright caught an early morning flight out of Fox Field in Lancaster and flew up to Redding.

In September, not long after Wright had extracted Guy from his ex-wife's house in Shingletown, Guy moved in with the Wrights. Guy had his mobile trailer transported from the Baldwin Park warehouse to the Wrights' property in Lancaster, located on the dry scrubs of the Mojave Desert on the outskirts of the Antelope Valley, north of Los Angeles.

At the Wrights', Guy had good days and bad ones—on some days, he could count on just a few good hours. As Dale remembered it, "You could tell he was going downhill." Some nights he spent just wandering around their house. The Wrights installed a second telephone line in their house for Guy because, as Dale explained, "he loved to talk on the phone." Sitting on their porch, Guy held forth about In-N-Out and drag racing and seemed excited about plans to launch his new Top Fuel team.

Then, on the morning of November 6, 1999, Guy disappeared. Dale remembered the day clearly because it was the wedding day of her eldest daughter, Janet. It wasn't the first time that Guy had gone missing—but this time they couldn't find him. At about 8:00 a.m., the Wrights called the police to report that Guy was gone. Sheriff's deputies found him wandering near a desert road and brought him to the hospital. When they located Guy, he was disheveled and wearing gray sweatpants and a white T-shirt with the In-N-Out logo.

Not quite a month later, at around midnight on December 4, the Wrights were sitting in their living room watching the TV Land channel with Guy when their eighteen-year-old daughter Darci noticed that Guy was slumped over in the chair where had been sitting. He had collapsed and was not breathing. "I remember thinking 'oh my God,'" said Dale. Her husband put a blood pressure cuff on Guy, but he had no pulse, and they called the paramedics. The emergency medical team arrived at about 12:30 a.m., but they found Guy unresponsive. Exactly two minutes after arriving at the Antelope Valley Hospital at 1:10 a.m., Guy Snyder was declared dead; he was forty-eight years old. Sheriff's deputies, who had arrived on the scene unaware of who Guy was, took note of his simple clothing and unexceptional motor home and wrote "Unemployed" on their report.

It was a tragic ending for a man for whom life had offered so many opportunities. Unable to measure up to his father's expectations, long in the shadow of his younger brother, unable to shake his substance abuse, Guy sank into a life of drugs and despair. Following Guy's death, the American and California flags and another with the words "Don't Tread on Me" flew at half-staff outside In-N-Out's Baldwin Park headquarters. "It was sad and kind of shocking when he died," said old friend Paul Althouse. "That boy had everything, and all that money." Later he wistfully exclaimed, "He was a hell of a guy. The drugs screwed up his life."

The Los Angeles County Coroner's office performed an autopsy on Guy at 10:30 a.m. on December 5, 1999. His body betrayed numerous medical problems and a history of drug abuse (specifically opiates and methadone). The attending coroner, Dr. Ogbonna Chinwah, found Guy's bloated body wracked with scars, bruises, and evidence of track marks on his arms. He had had an irregular heartbeat and an enlarged heart as well as active Hepatitis B and inactive C. The coroner declared that the cause of death was an accidental overdose of the drug hydrocodone, the synthetic opiate found in the painkiller Vicodin.

On February 6, 2000, two months after Guy died, the story of his arrest in Claremont three years earlier surfaced publicly. It appeared in a lengthy obituary that ran in the *Orange County Register* with the headline: "The CEO of In-N-Out Fought Drug Use and an Injury that Quashed His Racing Dreams."

Unlike the large public memorial given following the death of his younger brother, Rich Snyder, Guy's service was a strictly private affair. It was Lynda Snyder who made the arrangements for the family-only funeral; Guy was buried in his red In-N-Out racing jersey. Lynda's brother-in-law, an ordained minister, performed the service. A small memorial was held at the Pomona Raceway—among the attendees were Esther, Guy's racing team, his ex-wife, Lynda, their daughter Lynsi, and his former stepdaughters and their husbands. Wilbur Stites's widow, Kim, and their daughter Meredith were also there. It was the third time in twenty years that Esther had buried a son.

The Tuesday following Guy's death, In-N-Out's corporate lawyers called Jerry Darien to tell him they were pulling the plug on the Top Fuel team sponsorship.

There seemed to be a lot of blame and ill-feeling surrounding Guy's last days. Lynda and Mark Taylor excluded several friends and family members from the memorial. Many of Guy's friends, including those from his high school days, were barred. Significantly, the Wright family was also excluded. Kathy Touché said that she received a phone call from Taylor informing her that she was not welcome. "They didn't allow my folks to go either," she sniffed, "and they loved the hell out of him, it was cruel. Not much Christianity there."

At the time of Guy Snyder's death, In-N-Out Burger had grown to 140 stores, employing about sixty-five hundred employees. It was earning an estimated $212 million anually, an 8.7 percent increase from the previous year. Actually, in the six years between the time that Guy was named chairman following his brother Rich's death in 1993, the chain's revenue had jumped almost 83 percent. In-N-Out had pushed out geographically to Nevada and was about to move into Arizona. It had earned the unfettered loyalty of its customers, envy from its competitors, and plaudits in every imaginable category. When asked to characterize the mark that Guy Snyder left on the beloved chain, an insider remarked, "He just didn't change a thing. He knew well enough to just leave it alone."

CHAPTER 21

At seventy-nine, Esther Snyder stepped up to assume control of the chain that she had founded with her husband more than half a century earlier. "Don't let her age fool you," exclaimed Mark Taylor, who quickly began taking on a more prominent role in the company, to the *Los Angeles Times*. "She is as sharp as a tack. She is the hardest-working Snyder there ever was. She remains the president and is our leader and has all of our support."

Taylor's florid assurances did little to quash speculation about the future of In-N-Out. The chain that had for decades said little about itself was too often the subject of conjecture. And from the outside, at least, the company appeared vulnerable—just as it had six years earlier following the death of Rich Snyder. After all, the chain's sole heiress, Lynsi Snyder, was still in high school, just five months shy of her eighteenth birthday; Esther was fast approaching her ninth decade. Despite Esther's obvious determination and spirit, her old friend Carl Karcher voiced the concerns of many when he told the *Los Angeles Times*, "Esther is very, very tired. She's really had her ups and downs."

In the days following Guy's death, a group of experts with little to no connection to In-N-Out expressed theories as to what would happen next. Two camps emerged. The first raked over the complicated issue of succession. The second promoted the idea of a sell-off.

Bill Carlino, managing editor of the *Nation's Restaurant News*, stood with the latter faction. "I'm sure there would be no shortage of potential buyers," he remarked. "People would love to have a concept like that in their portfolio."

Historically, the statistics did not favor companies like In-N-Out Burger remaining intact. According to the Family Business Forum at the University of North Carolina, only 30 percent of all family businesses are successfully transferred to the second generation, while only 12 percent make it to the third. After that, less than 3 percent make a successful transition to the fourth generation. Moreover, usually a family business has several branches of descendants—siblings and others to whom the baton can be passed. At In-N-Out, there were only two family members remaining: an aging Esther Snyder and her teenaged granddaughter, Lynsi.

By the time that Guy Snyder died, family-owned burger joints like In-N-Out were a thing of the past. Only a very few had survived through the decades, the Apple Pan in West Los Angeles or Louis's Lunch in New Haven, Connecticut, among them. First opened as a lunch wagon in 1895, Louis's (among many others) claims to have invented the hamburger sandwich. Few had grown into a chain like In-N-Out while remaining both independent and iconic. Most of Harry and Esther Snyder's pioneering contemporaries had disappeared or were absorbed into the $107.1 billion fast-food industry.

The old-fashioned values that In-N-Out Burger represented were being resurrected in theme restaurants like Johnny Rockets, a successful chain of retro diners. Established in 1986 by the late Ronn Teitelbaum, a former clothing retailer, Johnny Rockets sold burgers and shakes in a re-creation of a postwar Southern California diner complete with tabletop jukeboxes, a U-Shaped counter, and padded booths. (The Apple Pan's Martha Gamble later claimed that Teitelbaum modeled Johnny Rockets after her family's West Los Angeles restaurant. Once, she said, she found him rummaging around in her garbage, checking out the brand of tuna they used.)

But in 1995, with about sixty shops, Teitelbaum sold his business to an investment group led by the New York venture capital firm

Patricof & Co. in a $35 million buyout deal. Twelve years later, Red Zone Capital Fund II (the private equity firm of Washington Redskins owner Daniel Snyder) purchased the chain for $126 million. By then there were 213 stores across the United States and overseas in Dubai, Mexico, and the Bahamas, as well as eight ships within the Royal Caribbean Cruise Lines. The new owners hoped to expand the chain further and build smaller, cheaper Johnny Rockets in places like airport terminals.

Sell-offs and franchising had been central elements of the fast-food game since the 1950s. More than forty years later, a new twist emerged, and all of the standard norms of business were being broken. The week that Guy Snyder died, the Los Angeles Times was dominated by news of the "new economy" propelled by technology. The newspaper proclaimed that the Internet was a "gold mine," and headlines trumpeted "Rally Heard Round the World, Dow Jones Industrial Average Skyrockets as Bull Market Continues," and "Strong Job, Pay Figures Fuel Stock Market Rise."

The game had changed. Up in Silicon Valley, a five-and-a-half-hour drive north of Baldwin Park, new paper millionaires were being minted by the busload. The idea of building a small company and watching it grow over the long haul had become passé. Everyone was looking for a quick fortune, cashing in and cashing out. Company loyalty was a longtime thing of the past. Employees were trading jobs like baseball cards, hiring themselves out to the highest bidder in exchange for a drawer full of stock options.

In-N-Out Burger had long operated in its own orbit. A more-than-capable bench of managers led the chain; dedicated to the principles outlined by the Snyders, many of them had been at the company for decades. The immediate transition was neither as troubled nor uncertain as some might have expected (or even hoped). As Carl Van Fleet, the company's vice president of operations, told the Los Angeles Times, "We have a team of people out in our stores that are all similarly committed, and I think that goes a real, real long way and speaks to how we have gotten through the tragedies." With the present taken care of, it was the future that everyone else thought was up for grabs.

Inside the company's headquarters, grief and recriminations filled the air. Less than a week after Guy's death, Tom Wright was told that his services were no longer required. Wright was asked to resign from the company, the board, and as a co-successor-trustee of the family trusts, and was given a severance package.

Guy's death put the successor order over the family trusts into effect. Tom Wright was off the list, and soon thereafter Doug Ammerman resigned as a cotrustee as well. According to one account, the reason given was potential conflict of interest (Ammerman later went on to serve on the boards of Carl Karcher Enterprises and the casual clothing company Quicksilver). That left Richard Boyd. With Wright and Ammerman out, Mark Taylor was installed. The two men became trust cotrustees as well as co-executors of Guy Snyder's estate.

Boyd and Taylor were responsible for administering considerable assets, including nearly two-thirds of the company's private stock, for the eventual inheritance of Lynsi Snyder. Under the terms of both the Esther L. Snyder Trust and the Lynsi L. Snyder Trust, Lynsi would not begin to receive shares of each of the trusts until she turned twenty-five, and then only on a staggered basis: the first third of the shares would be dispersed upon Lynsi's twenty-fifth birthday; at thirty she would receive one-half of the remaining shares; and at thirty-five years old, she would receive the balance of the shares. Under the terms of the trusts, as cotrustees, Boyd and Taylor each had the power to vote one-half of the stock. It wasn't long before the shared responsibilities between Boyd and Taylor led to an exchange of legal blows.

When her father died, Lynsi also became the sole beneficiary of the Harry Guy Snyder Testamentary Trust. By 2005, the trust's assets were worth in excess of $12 million. Its primary assets were a 70 percent interest in Snyder Leasing and a 41.67 percent interest in Snyder Properties, a California general partnership that owned a commercial building and land in Riverside as well as interests in a string of residential and commercial properties in California, as well as 80 shares of common stock in the Garden Grove In-N-Out Burger and 1,175 shares of common stock in the In-N-Out store in Fontana. Lynsi's inheritance also included cash, various properties, shares of stock in several security instruments, 1,000 shares in Flying Dutchman Racing Inc., and her father's extensive collection of cars and other vehicles (Guy put in a stipulation that his daughter not receive any of his Porsches until she reached the age of twenty-one). Lynsi was entitled to a third of this trust upon turning thirty, an additional third at thirty-five, and the remainder upon turning forty.

Guy's thirteen-page last will and testament, signed on May 18, 1999 (less than seven months before his death), left his first wife, Lynda, his former stepdaughters, and their husbands (including Mark Taylor) with nothing. Except for provisions naming Lynda Snyder guardian of the estate of their minor daughter, Lynsi, and allowing for financial "expenditures necessary for the benefit of the minor child," his ex-wife and her family were effectively cut out. "He did not leave them a dime," explained one friend. "He felt they turned their backs on him."

Guy did bequeath gifts to a select few. He gave cash totaling $43,000 in varying amounts to a group of four friends, and his Harley-Davidson motorcycles to one of them. Tom Wright received Guy's Glock pistol, a Desert Eagle .45 caliber pistol, and $50,000 in cash. Guy left Tom and Dale Wright's daughters Robin and Darci joint ownership of his Pontiac. However, according to Dale, Lynda stepped in and said she still had title to the Pontiac and refused to relinquish the car to her nieces.

Perhaps underlining just how close the two men had become, earlier Guy had named Wright as the executor and trustee of his estate.

All eyes were on Esther. Since breaking her hip while in Redding for the store opening and an extended visit with her granddaughter, Lynsi, about two years earlier, Esther had been in ill health. The break required at least two surgeries and physical therapy. She ended up with an infection that left her severely weakened, and she was placed in a convalescence home in Northern California where she stayed for the better part of a year. She remained fragile, often needing the assistance of a cane or a walker, and at times she used a wheelchair to get around. Esther's movements had become painfully labored. Years earlier she had undergone heart surgery, and now a host of other health problems had developed.

As a result, her daily routine was curtailed significantly. No longer able to drive each weekend to Costa Mesa to attend church services at the Calvary Chapel, she often listened to the pastor's sermons on audio and videotapes. She now needed a driver to chauffeur her to the company's headquarters in Baldwin Park or Irvine. Often she worked out of her Glendora home, a ranch house shaded by oak trees and fronted by a white fence and a mailbox that was a miniature version of the main house. There she maintained a vigilant oversight over the company's financial affairs and tended to a host of civic and philanthropic activities (including the In-N-Out Foundation).

Frequently housebound, Esther was consulted by managers on company matters. Board meetings were conducted in her living room or via phone. A crew from *Burger Television* traveled to Glendora. She gave rallying speeches to be shown to the chain's associates. Weakened but not incapacitated, Esther insisted on being involved in all of the chain's decisions. Whenever she could, Esther continued to visit the stores, ordering a burger through the drive-through and taking it apart to make sure the bun was properly toasted and the tomatoes placed correctly (and that the associates were smiling and friendly).

Owing to her failing health, in February 2000, Esther bowed out of attending the company's annual black-tie dinner at the Disneyland Hotel. Each year she addressed the associates at the big fete and that year, although she wasn't present, Esther gave her speech through a speakerphone. Transcendently popular, the group gave her a thunderous standing ovation. Several months after the dinner, Caroline Haley, a fifteen-year associate and manager of the In-N-Out near the Los Angeles International Airport, told the *Los Angeles Times*, "She's wonderful. She's so into the people of the company." Haley—who shocked her parents when she quit college to move up the ladder at the burger chain—explained, "It really goes back to her and her husband's commitment to take care of the people."

A private woman, Esther rarely spoke of the heartbreaking loss of her husband and her sons. A gentle and gracious soul, it was her faith that kept her steady through all of the emotional upheavals, and she refused to dwell on the difficulties or losses that plagued her. Meredith Stites, who grew up as Esther's granddaughter, called her an incredibly humble woman. "Grandma was always very loving, so happy and grateful for everything," she said. "She lost everybody in her family, but she still believed in God. She had all those things happen, but she had such a great outlook on life."

Optimistic by nature, Esther, with her positive outlook, was a comfort to those around her. She was always expressing her gratitude for what the Lord had given her. "She thanked the good Lord that she still had her eyes to read her books," her friend Wolf Kahles remembered her saying often.

Kahles, a German businessman from Bietigheim-Bissingen, a town fourteen miles north of Stuttgart, first met Esther in 1999 through a mutual friend. At the time, he owned Kentucky Fried Chicken and Pizza Hut franchises in Germany. The meeting took place at her home. She was still recovering from her hip surgery and needed a walker, but, he recalled, "she opened the door herself, there was no maid." And he remembered, "She had all of her business papers on the dining room table and she was working there."

With Kahles's son studying at a private boys' boarding high school in Claremont, he made a point of calling on Esther whenever he came to visit. The two developed a friendship. "She was this beautiful, fantastic human being," he said. She was very proud of In-N-Out. "It was her business that she built up with her husband," he said. "She looked after every detail. It wasn't about the money for her, but about her associates in the business being happy."

According to Kahles, the two often talked about the fast-food industry. Kahles had soured on the parent corporation of his franchises and later sold his interests in them. "'Don't go with a large corporation,' she advised. She always told me, 'Do it yourself and do it right, believe in the Lord and everyone will be happy.'"

When Kahles asked about expanding In-N-Out nationally and even approached her about taking the concept to Germany, Esther demurred, telling him that she wanted the company to grow, but slowly—first in California, then in Nevada, and then the surrounding states. "She never wanted the business to explode but to grow naturally by demand," he recalled. "If the customers and associates were happy and coming back, that was how they grew the business."

Esther was aware that the company was expanding further, and it was a concern. She told her nephew Joe Stannard that she thought the chain's growth might be moving too fast and that In-N-Out's quality and service might suffer. There seemed to be other voices that had a few ideas about how far and how fast the chain could go. As emphatic as Esther was about keeping In-N-Out independent and family-owned, she was also pragmatic; she wouldn't be around forever. For several years, friends and observers had noted a subtle sense of anxiety in her when it came to issues surrounding the chain's future.

It was Esther's intention that Lynsi eventually take over the business if she so desired. A strong advocate for education, while Esther wanted Lynsi to run In-N-Out, she also hoped that her granddaughter would get a college education first.

In-N-Out heiress Lynsi Snyder was born into a world of wealth and privilege. The only surviving direct descendant of Harry and Esther Snyder, the future of the entire hamburger empire was dropped onto her slim shoulders. Lynsi had her grandmother's round, wide face and brown, almond-shaped eyes. The only natural grandchild, she was Esther's pride and joy.

When Lynsi was born, her two older half-sisters, Terri and Traci Perkins, were already in their teens. Theirs was a knotty set of family ties with overlapping bloodlines and bank accounts. In line to inherit hundreds of millions of dollars, the family's money was tied up in In-N-Out Burger, and Lynsi was the one and only access point to it.

While living in Glendora, she was said to have chafed at the taunts of the other schoolchildren, who teased her after finding out about her connection to In-N-Out Burger. They called her the "Burger Princess." As someone who knew her at the time explained, "she didn't really enjoy being the brunt of jokes. When they found out who she was, the kids would prejudge her before she made friends."

By the time that Lynsi was about twelve years old, the teasing stopped. The family had been living in Shingletown on the Flying Dutchman Ranch. It was a small town with a population of just over two thousand people. There Guy and Lynda built a private, non-denominational Christian school for their daughter. It was actually more of an elaborate home school; the Snyders bought a house and property in the area and filled the house with desks, computers, books, and perhaps ten local students. They hired a teacher and devised a curriculum based on Christian principles. The school remained in operation until Lynsi graduated from high school.

Easygoing, but with a sharp sense of humor, Lynsi was said to crave attention and liked talking about her ideas and thoughts. She was also creative and had an artistic side. Like her father, she got the drag racing bug early; strong willed, she also shared her father's stubborn streak and his passion. Lynsi grew up sheltered and out of the public eye.

Despite his own considerable problems, Guy was quite clear when it came to his hopes for his daughter's future. According to friends, he wanted Lynsi to learn the business from the ground up like he and his brother had before taking it over someday, but not until she was "mature and had sufficient experience to successfully manage the company." He was said to have concerns about his daughter becoming spoiled as a result of the immense fortune she was born into. As one intimate put it, "She never went without anything." Like his mother, Guy wanted Lynsi to go to college. He wanted her to make a success of her inheritance. Presumably, that is why the family trusts were structured so that Lynsi would not obtain enough stock to control In-N-Out until she reached her thirtieth birthday.

Like many teenagers, Lynsi liked to dance, work out, and go to rock concerts. She had a part-time job, too—for a time, Lynsi worked at the In-N-Out Burger in Redding. Unlike most teens, however, she married right after high school, at the age of eighteen. Lynsi's husband was a local boy named Jeremiah Seawell, who was not much older. The couple had met about four years earlier on a snowboarding trip. Seawell was said to have been besotted with Lynsi; before he died, Guy told friends that he thought the boy followed Lynsi around like a "puppy dog."

The wedding took place in the summer of 2000. It was a young girl's idea of a fairytale. The ceremony was in a Pasadena church; the reception for two hundred guests was held at a nearby mansion. There was a horse-drawn carriage; butterflies and doves were released into the sky while overhead a skywriting plane spelled out "I Love You."

There were several rumored reasons for the young wedding. In one version of the tale, it was the classic escape from a conservative upbringing. Although Guy was said to have opposed the nuptials on that grounds that Lynsi was too young, in another version, it was his ex-wife Lynda who pushed it forward.

After Lynda and Guy's divorce, Lynda maintained what several insiders described as a very strong interest in the affairs of the Snyders and In-N-Out Burger. In fact, it was Lynda who telephoned Esther to inform her that Guy had died. At the time, Lynda was said to have become deeply involved in the Successful Christian Living International Association of Ministers and Churches based in Redding, California. The church's founder Steven A. Radich, a self-proclaimed "Apostle," described himself as "called by God while still in his mother's womb, to bring an end-times message of healing and transformation for all the nations." Its mission was to "fight the good fight of faith, engag[e] in spiritual warfare against the five enemies of the faith." In addition to what it called "mere rationalistic Christian teaching," the church advocated a more confrontational "high touch" approach that included "personal salvation and personal deliverance from demonic influence." By most accounts, Lynsi was heavily influenced by her mother. Another theory emerged concerning the young marriage; it was Lynda's bid to keep a hand on the In-N-Out wealth. While Lynsi was the sole beneficiary of the Snyder family trusts, in the event that she died without leaving any living descendants (and without exercising her testamentary general power of appointment), all assets of trusts—including all of the shares of In-N-Out stock that had not been previously distributed to Lynsi—would be transferred to the living siblings of Esther and Harry Snyder and their children. According to a stipulation of the trust, Lynsi's mother Lynda, her two half sisters, and any husband of hers would receive absolutely nothing.

At Lynsi and Jeremiah's wedding reception, the bride's lengthiest dance was with her mother. Perhaps it was a sign; the marriage did not last long. Within two years the couple separated and quietly divorced.

In 2005, Lynsi married Richard Martinez in a small ceremony in San Diego. Martinez worked for In-N-Out dealing with the chain's cookout trailers, and the two were introduced when they were partnered to learn and perform dance routines at the annual In-N-Out associates' picnic. The couple later moved into a custom-built $1.21 million four-bedroom house on Valiant Street* in a McMansion-filled cul-de-sac in Glendora. Fronted by perfectly manicured lawns and flanked by nine graceful palm trees, the cream-colored house with a gray slate tile roof and paned windows was situated among the foothills of the San Gabriel Mountains. The house, just minutes away from Esther's, had a swimming pool and spa, a detached guesthouse, and a three-car garage.

Soon, the couple were said to have become deeply involved in the Successful Christian Living Church. They were said to hold prayer- and church-based counseling sessions in their home for groups of twenty to thirty church members. Before long, according to intimates, Lynsi's new husband, Richard Martinez, became a minister.

With her father gone, expectations were that Lynsi would learn the In-N-Out ropes like Guy and Rich had done, in anticipation of

---

* According to the Office of the Assessor of Los Angeles County, the house was sold in March 2007 for $1.215 million

her considerable inheritance. She spent some time working in different departments, rotating between human resources and the meat department, and Lynsi was made merchandise manager of the Company Store. She helped design T-shirts and catalogs and created items for the managers' promotional trips. It was a full-time job, but Lynsi worked when she wanted to, as one former colleague and intimate put it, "probably because she knew she didn't have to. She was dedicated and responsible, but on her own terms. If there was a meeting, she'd show up; if there were no meetings, she didn't."

Like many heirs, Lynsi's inheritance was a mixed blessing. As the intimate explained, "She seemed overwhelmed with the responsibility, but looking forward to taking it over and making the adjustment." At the same time, she displayed a certain ambivalence about her future. "I think most of us choose what we want to do," said the intimate. "This is what she was born into. It was a role she had to accept. I don't know if she wanted it."

CHAPTER 22

Everyday life at In-N-Out Burger continued. By 2000, the chain had grown to 142 stores, while revenue had climbed to an estimated $220 million. The numbers, however, were only a small part of In-N-Out's story. For some time—actually quite some time—the chain had moved well beyond its humble beginnings as a burger joint. Without dispute, In-N-Out was now elevated to cult status.

In June 2000, the thousands of spectators who had descended upon Pomona for the legendary So-Cal Speed Shop's annual hot rod open nearly went ballistic; So-Cal had missed the six- to nine-month advance deadline to order one of In-N-Out's cookout trailers. "All we got were complaints," moaned Tony Thacker, who was So-Cal's marketing director at the time. "There were five thousand people, many who came from Australia, Japan, and Europe, and for all of them, In-N-Out is part of the whole experience." Thacker, who later became director of the National Hot Rod Association Museum, exclaimed "We almost had a riot on our hands."

In-N-Out was not just a place where people wanted to eat—it was a company where people wanted to work. When in 2002 In-N-Out opened a new drive-through in the coastal town of Oxnard (a forty-minute drive south of Santa Barbara), there were nine hundred applicants for seventy positions.

For the rest of the fast-food industry, however, the turn of the millennium left a bad taste. A great paradox was unfolding. The fast-food hamburger was more popular than ever. Fast-food sales in the United States, well on their way to approaching $150 billion, certainly attested to that. In the century since the hamburger had first been served at the St. Louis World's Fair in 1904, it had gone from the food of the poor to America's national meal. However, it was also the most vilified food in the world—McDonald's, which was at one time considered such an American icon that in 1971 Norman Rockwell illustrated its annual report, now represented for many the nation's darkest impulses. Anti-globalization protesters attacked golden arches all over the world as an emblem for American imperialism. At home, the industry's reputation didn't fare much better. In 2002, fast food ranked dead last on the University of Michigan's American Customer Satisfaction Index—lower than even the Internal Revenue Service.

Perhaps it should come as no surprise, then, that during this time the perceived all-encompassing evils of the industry on society and humanity were cataloged in a pair of media sensations. First, Eric Schlosser's best-selling *Fast Food Nation* was published in 2001. Three years later, Morgan Spurlock released his documentary *Super Size Me*. In the film, Spurlock eats only McDonald's during a thirty-day period and chronicles a litany of physical (heart palpitations, weight gain, high blood pressure) and psychological effects that wreak havoc on his body as a result. Schlosser's book rattled the country; Spurlock's film, which earned an Academy Award nomination, repulsed it. Both sent a chill down the spine of fast-food executives everywhere. Outbreaks of *E. coli* bacteria and mad cow disease as well as connections to obesity further eroded fast food's standing in the public imagination.

The industry was battered. In January 2003, McDonald's posted its first ever quarterly loss—$343.8 million—since it went public in

1965.* As the chain grew to more than thirty thousand stores spanning the globe, service and quality had slipped. The fast-food giant, which constantly churned out new products in an effort to goose sales, hadn't produced a blockbuster since introducing Chicken McNuggets in 1983. It appeared that McDonald's was running out of steam. In April 2003, *Fortune* asked "Can McDonald's Cook Again?"

McDonald's wasn't alone. The industry that had experienced ceaseless growth for decades was in a ditch. Its reputation was badly tarnished. Health woes, market saturation, dirty restaurants, and surly service led customers to abandon fast food in droves. At the same time, the idea of eating fresh, local, and additive-free food was gaining currency. A new market segment called "fast casual" that offered upscale sandwiches began siphoning off consumers from fast-food restaurants. Over the next few years, the big chains aggressively attempted to combat their image as junk food peddlers and boost sales.

At McDonald's, newly installed CEO Jim Cantalupo oversaw the company's comeback strategy, aptly named the "Plan to Win." Over the next several years, the world's largest restaurant company rolled out a slew of new initiatives to boost profits and attract a new breed of customers. New tactics included opening fewer stores and improving service, extending hours of operation, and emphasizing breakfast. In 2002, McDonald's established its "Dollar Menu." The value-priced Big Macs and Yogurt Parfaits helped to fuel a 33 percent sales spike over the next three years.

In a truly bold move, McDonald's poached four-star chef Dan Coudreaut from the Four Seasons Hotel to help revamp its menu. The company had launched a series of lighter and healthy menu options, too. The fast-food giant sold premium coffee and chicken wraps at home while pushing even further into foreign markets. In 2004, the company announced a complete store redesign to encompass its res-

---

* During the previous eleven months, McDonald's had nothing but consecutive months of negative same-store sales.

taurants worldwide: out with the plastic seats and in with the clean, modern lines, muted colors, brick and wood, modern light fixtures, comfortable armchairs, sofas, and wi-fi connections.

While the burger barons went back to their test kitchens, turned to focus groups, and undercut each other on price, In-N-Out remained the same and continued to grow. In 2003, sales reached an estimated $302 million, up 9.8 percent from the previous year. Even with the turmoil within the industry and an unenthusiastic IPO market, industry watchers still placed their bets on In-N-Out, believing that the chain had the ability to raise a great deal of capital if it decided to go public.

In-N-Out was not, however, totally insulated from the problems plaguing the industry. In May 2003, the company quietly settled a lawsuit filed by the family of a Palo Alto girl who claimed she contracted *E. coli* after eating at In-N-Out's Kettleman City store off the I-5 between Los Angeles and San Francisco. The incident seemed to be an aberration on an otherwise spotless record.

The Kettleman City episode occurred the same year that the editors of a Los Angeles–based monthly online magazine, *Vegetarians In Paradise*, staged a protest against In-N-Out's "Food for Thought," program. Esther remained a booster of reading, and the literacy program, in conjunction with public libraries, gave students hamburger coupons and Achievement Award certificates in exchange for having read five books. In February, editors Zel and Reuben Allen launched a letter-writing campaign—"Say No to In-N-Out Burger"—targeting 126 libraries in those states where In-N-Out operated.

The Allens' remonstration was not the rousing success that the vegetarian activists had hoped it would be. In their April 2, 2003, newsletter, the pair reported that their earnest protest was "greeted with a giant wave of indifference." Among the few responses that *Vegetarians In Paradise* did receive was one from Brian Lewis, the county librarian for Tulare County in California, who described the illiteracy rate among Tulare adults as more of a problem than the eating habits of its youth. "Becoming literate may help our youth learn about healthy eating," he wrote. Lewis and the head librarian for the city of Oxnard pointed out the secret menu to the Allens, explaining that In-N-Out would make a vegetarian or even a vegan version of their hamburgers upon request.

And yet in the midst of what many had predicted might augur the collapse of fast food and with it the end of the hamburger's exalted place as part of America's culinary heritage, the game changed once again. It happened during the hamburger's darkest hour and at the hands of a foreign national. In 2001, nearly six months after the publication of *Fast Food Nation*, Michelin-starred chef and New York restaurateur Daniel Boulud introduced what came to be known as the gourmet burger. Asked to comment on the protests against McDonald's in his native France (which raged in spite of McDonald's huge success there), Boulud told the *New York Times*: "The French are jealous . . . they wish they could have invented McDonald's." Soon after, Boulud invented the twenty-nine-dollar DB Bistro Burger at his newly opened DB Bistro Moderne on West Forty-fourth Street.

Made with ground sirloin stuffed with tender red wine braised short ribs, foie gras, a mirepoix of root vegetables, and preserved black truffle, the burger was served on a homemade toasted parmesan and poppy seed bun spread with a touch of fresh horseradish, oven-roasted tomato confit, fresh tomato, red onions, and frisée lettuce. Boulud's hamburger quickly became the most talked-about dish of the year. The *Atlanta Journal-Constitution* called Boulud's entry the "debut of the gourmet hamburger." In 2003, Boulud added shaved black truffles to the DB Burger and introduced the sixty-nine-dollar "Royale."* Soon thereafter, chefs all over the country introduced their own versions of the gourmet burger. The Old Homestead, reputedly the oldest steakhouse in Manhattan, introduced its twenty-ounce, one-hundred-dollar Tri-Burger made from Japanese, American, and Argentinean beef at its Boca Raton location. In Minneapolis, Vincent

---

* In 2007, Boulud announced the opening of a new New York casual brasserie featuring, of course, his signature hamburger.

stuffed burgers with short ribs and smoked Gouda. Before long, rumors circulated about a restaurateur who was said to gild his burger buns with twenty-three-karat gold leaf.

The organic movement's mantra of fresh, local, and seasonal ingredients was being applied to that most prosaic American dish: the hamburger. New burger places began sprouting up touting quality ingredients and custom orders. Places like Father's Office in Los Angeles and the Shake Shack and Five Guys in New York catered to boomers and others with sophisticated palates and well-padded wallets. They seemed to trade on a growing number of foodie movements while tracing their lineage back to the postwar American food revolution that spread fast food across the country. They were often described as a kind of aspirational In-N-Out Burger. It wasn't long before the big chains introduced their own versions of gourmet burgers. And the French came to embrace the hamburger too; all over Paris, Michelin-starred chefs translated the hamburger into their own gastronomy.

Sitting in DB Moderne, where he had made burger history, Boulud waxed philosophical about the vaunted burger's reputation. "The strange thing about burgers is they are so plain and bland and yet so good," he said. Boulud, who recalled his first burger as a fourteen-year-old boy in Lyon, said that he was not the least surprised that a burger could inspire a cult. "People care about burgers, and they will drive miles for the right burger. They talk about burgers like fanatics." According to Boulud, DB Moderne sold about one thousand of his burgers each week. "We have customers flying out on private jets to order the DB Burger."

Several years ago, a friend took Boulud to an In-N-Out in Los Angeles. He was impressed. "They weren't too big, and I like that I can hold them in one hand," he explained. "Americans understand that people live in their cars; as long as you can drive and hold a meal in your hand, you will be a success." In-N-Out, he said, inspired him. "People like me are always looking for the icon of perfection, but the success benchmark doesn't necessarily have to be the Four Seasons Hotel. It doesn't matter. There will always be successful companies that have kept their vision of quality. In-N-Out is not about size. What they have done sends a message that they are concerned about the product." He joked, "Maybe I should do a trade and switch with In-N-Out—I might learn something."

Daniel Boulud wasn't the only gourmet talking up In-N-Out Burger; in a world in which Michelin chefs were flipping foie gras–stuffed hamburgers, In-N-Out stood in a class all by itself. Ruth Reichl, the former restaurant critic of the *New York Times* (who at one time lived in Los Angeles), told the *Times* that In-N-Out was "a great California institution." It was, she noted, a point of pride.

The influence of the little Baldwin Park burger joint had spread far and wide. To say that Michelin-starred chef Thomas Keller was an In-N-Out fan is an understatement. "I probably have thousands worth of gift cards, hats and pins and T-shirts," proclaimed the man whose gourmet restaurants—The French Laundry in Yountville, California, and Per Se in Manhattan—have earned him the sobriquet of America's best chef by nearly every respectable gourmet. "People are always giving me In-N-Out things. They know how much I like it." Almost fifteen years after he was introduced to In-N-Out, Keller could still recall his first visit to the little burger chain. It was the summer of 1991, and his friend David Lieberman, a Los Angeles–based artist's representative and wine aficionado, made the introduction. "I had told him that I was a fan of hamburgers, and he said, 'I have to take you to In-N-Out. You have to try it.'" The pair went to the West Los Angeles location on Venice Boulevard and made a meal of cheeseburgers and a bottle of Ridge Lytton Springs Zinfandel. It became a routine; when Keller was in Los Angeles, the two friends would pick a bottle of wine and hit an In-N-Out.

Like Boulud, Keller—whose innovative, intensely flavored menus have been likened to the pleasure of eating food with a "sprinkling of magic dust"—is both inspired and dazzled by In-N-Out. "It amazes me that the company has stayed focused," he explained. "And sure, it's a hamburger place, but while the other chains continue to evolve, it's become part of a trend. The others are always looking for the next trend." Although In-N-Out's most expensive item is its $2.75 Double-Double whereas French Laundry serves a $240 prix fixe menu, Keller insisted that the two establishments were more similar than not. "If you think about cooking, you'll find at In-N-Out or French Laundry, it's about product and execution that's consistent. The quality of a restaurant goes back to the quality of the product and its execution."

Perhaps Keller's greatest valentine to In-N-Out was his announcement in 2006 that he was opening up his own burger joint across the street from his French Laundry in Napa Valley to be called Burgers and Half-Bottles and selling just that—hamburgers and wine. "A really good hamburger is part of American culture," he explained. "It will be a high-quality hamburger, much the way In-N-Out does it. What better way to express the true meaning of simplicity?"

Indeed, as the industry worked to claw its way back from an abyss of lawsuits, health scares, quarterly losses, and a revolving door of executives and business strategies, In-N-Out's cult status was being solidified in ways both great and small. That in-the-know feeling elicited by the chain was gaining traction with a wider audience. West Coast rappers made reference to the chain in their songs. "Did just that at In-N-Out Burger / No pickles, no onions, no playin'," rapped Andre Nickatina in "Cadillac Girl." In 1998, a wry In-N-Out reference found its way in the Coen brothers' cult hit *The Big Lebowski*. The dialogue between Walter Sobchak, Donny, and the character simply known as the Dude goes as follows:

WALTER SOBCHAK: He lives in North Hollywood on Radford,
    near the In-and-Out Burger—
THE DUDE: The In-and-Out Burger is on Camrose.
WALTER SOBCHAK: Near the In-and-Out Burger—
DONNY: Those are good burgers, Walter.
WALTER SOBCHAK: Shut the fuck up, Donny. . . .

Then in 2001, Graydon Carter, the editor of *Vanity Fair*, began hiring one of In-N-Out's cookout trailers for the glossy magazine's annual Oscar party. It began almost as an afterthought. "We started serving In-N-Out burgers late in the evening of the Oscar party," explained Carter, who said that he ate his first In-N-Out sometime during the 1970s. "The restaurants have a point to them, and they have stayed consistently on message. Plus, they make a mean hamburger," he said. Soon, gossamer-draped and tuxedoed stars were photographed tucking into cheeseburgers. Moreover, celebrities talked about the chain to the media even as they discussed their designer attire. On the red carpet at the awards show in 2004, actress Jennifer Garner told comedian Joan Rivers that after the ceremony she was heading straight to an In-N-Out Burger. The little burger chain became a part of the evening, like the glamorous after-parties. With little prodding or formal arrangement, In-N-Out had developed a broad base of very influential, adoring, and devoted fans.

One of the most extreme (if not bizarre) instances of adoration began in 1999 when a sixty-two-year-old Texas businessman named James Van Blaricum hatched a secret plan to own a piece of In-N-Out Burger in the Fort Worth area (reportedly after the chain refused to entertain his earlier offers to either invest in or franchise the business). In-N-Out was hardly amused by the venture. It was not merely an homage, they contended, but a complete copycat. Van Blaricum apparently went to such incredible lengths to duplicate In-N-Out Burger that he eventually landed in court, accused of conducting an "elaborate industrial espionage scheme." After opening the first of his planned one thousand Lightning Burgers stores at the Six Flags Mall in Arlington, Texas, in October 2001, In-N-Out Burger slapped him with a pair of lawsuits in Dallas and Los Angeles federal courts alleging trademark violation.

The plan, eventually laid out in court documents and news reports, resembled the plot of a dime-store potboiler, complete with moles and secret laboratory samples. Van Blaricum allegedly set out to reproduce In-N-Out's extraordinary success by copying everything from the diameter of the chain's hamburger buns to the blueprint of its kitchens. A group of eight employees (some of whom were former In-N-Out associates) headed by Jason Newling, his vice president of operations (later turned whistle-blower), worked for various In-N-Outs while on Van Blaricum's payroll. Samples of the hamburgers were frozen in sealed kyro-vac containers and sent to a lab for analysis (which would allow them to be duplicated). While undercover, the group reportedly was instructed to steal everything that wasn't nailed down and digitally photograph the rest.

The suit was settled on February 14, 2002. Van Blaricum was required to cease operations, transfer all assets of Lightning Burgers to In-N-Out, pay $250,000, and was prohibited from selling hamburgers for ten years. The seven defendants/moles (not including whistle-blower Newling) agreed to pay $250 each and were barred from working for any business associated with Lightning Burgers.

Van Blaricum, however, remained defiant. In April 2002, he denied any wrongdoing to the *Fort Worth Star-Telegram*, saying that In-N-Out "couldn't prove anything. I personally froze some meat and brought it back [to Texas]. And I did ask a company if they could evaluate it," he explained. "But I told them we do not want to duplicate anything." Clearly embittered by the whole experience, he fumed to the paper that he had spent "hundreds of thousands of dollars to make sure we didn't copy anything," only to have "nothing to show for it." Suffering from heart disease, Van Blaricum proclaimed that he was through with the burger business and announced that he had even become a vegetarian. "I'm not even going to eat hamburgers," he declared.

Through the years, as In-N-Out's success continued, its reputation spread as well, further and wider than the chain itself. The scarcity of In-N-Out stores seemed to invite a host of imitators, making it especially vulnerable in the forty-seven states where it was not located. It was a situation that did not go unnoticed on the top floors of the University Tower in Irvine. Led by the company's longtime general counsel Arnold Wensinger, In-N-Out embarked on a fierce policing strategy to protect what it viewed as its most important asset: its reputation. In particular, In-N-Out went after any and all entities that appeared to infringe upon its trademarks, hiring private investigators to monitor and gather evidence, firing off letters pressing violators to change their names, and eventually filing lawsuits. Frequently, their targets were businesses that had nothing to do with burgers such as Chinese restaurants, auto parts shops, and video game stores in areas where the chain wasn't even located.

A year before In-N-Out battled Van Blaricum, it was involved in a legal dispute with Texas-based Whataburger. In December 2000, In-N-Out Burger filed a federal lawsuit against the five-hundred-unit chain (located in ten southern states), accusing it of violating the Federal Trademark Act by using the words "Double-Double." Alleging that Whataburger's use "diminished and devalued the worth of the Double-Double mark and In-N-Out's business reputation and goodwill with the public," the Irvine chain asked for $75,000 in damages as well as reimbursement of legal fees and other costs. The two parties reached a settlement, and Whataburger (founded in 1950) stopped calling its sandwiches "Double-Doubles."

In-N-Out took its trademarks very seriously. The chain claimed to have used the "Double-Double" phrase since 1963 and registered it with the U.S. Patent and Trademark Office in 1981 (in 1997, it registered the Spanish translation *Doble-Doble*). Although "Animal Style," "Protein Burger," "3x3 Burger," and "4x4 Burger" had long been part of the secret menu, it was no secret that In-N-Out viewed the terms as proprietary, and the chain federally registered those trademarks too. In-N-Out meant business.

Jerry and Margie Rizza, a Eugene, Oregon, couple, found that out the hard way after In-N-Out sued them in September 2003, days after the Rizzas opened an In & Go Burger with a white, red, and yellow color scheme. The lawsuit, which claimed that the Rizzas had purposefully "appropriat[ed] and trade[d] upon In-N-Out's extensive goodwill," demanded that the Rizzas cease using the In & Go name, and required that they turn over any profits they made while using it. Rather than fight it out in court, the Rizzas settled with In-N-Out. They agreed to modify their color scheme and remove the words "In" and "Go," leaving the restaurant with the name & Burger.

Then, in May 2007, a new burger drive-through called Chadder's opened up near the local Applebee's in American Fork, Utah (population 22,387). Until Chadder's opened for business, American Fork had been known mostly for its marching band. Undefeated as state champions since 1990, the band had done American Fork proud, having performed at the Rose Bowl as well as the Macy's Thanksgiving Day Parade. But shortly after Chadder's opened, In-N-Out began receiving inquiries from customers asking whether the burger place was In-N-Out's new Utah chain. After all, Chadder's building exterior, color scheme, signage, menu, and employee outfits were all deceptively similar to In-N-Out's.

Perhaps adding to the confusion, for the past couple of years, news had spread that an In-N-Out was going to open sometime in 2007 in Washington, Utah; it was to be the first In-N-Out in Utah and the first outside of the chain's established three-state radius. With the Las Vegas distribution center already in operation, it made economic sense to move as far as possible while remaining able to deliver its fresh meat and ingredients daily. A location had been chosen on Telegraph Street, according to the city's mayor Terril Clove, and a pair of crossed palms had even been planted. "Thousands of people can't wait until they get here," Clove exclaimed in February, three months before Chadder's had opened. "The most common question I get as mayor is, 'When is In-N-Out coming?'"

Piqued by the reports of similarities between the two restaurants, in June, Arnold Wensinger was dispatched to American Fork. Once there, In-N-Out's attorney stood in line and ordered an "Animal Style Double-Double with Animal fries" and was served his meal In-N-Out style, the burger partially wrapped in paper and served in a cardboard box. Not long after Wensinger's visit, In-N-Out filed a thirty-page trademark infringement suit against the restaurant and owner Chad Stubbs in Salt Lake City and sought to shut down Chadder's as well as recover monetary damages.

News of the lawsuit spread quickly. The local media descended upon Chadder's, which actually boosted business for a time. Once the lawsuit had been filed, Chadder's, which had denied the allegations, began changing its look. Employees' aprons and hats went from red to blue, the background color of its menu board went from white to yellow, and the font on its menu board was changed. In July, a month after the lawsuit had been filed, U.S. District Judge Ted Stewart ruled that Chadder's could stay open but that it could not sell, advertise, or fill any orders using In-N-Out Burger's trademarked names. Should a customer ask for an Animal Style, Protein Style, a 3x3, a 4x4, or a Double-Double, all Chadder's employees must "say that the restaurant doesn't offer those items and refer them to Chadder's menu."

While In-N-Out was not able to put Chadder's out of business, the court's decision did put other imitators on notice. As far back as 1998, the year that In-N-Out went after five small businesses, Arnold Wensinger told the *Los Angeles Times*, "We don't want to be known as a bully, especially in the legal field. We're just trying to protect the rights to our name." However, he added, "They think they can ride on the coattails of an existing company. Why should somebody be able to use the name at our expense?" In-N-Out had proved to be a small chain with a big legal bite.

CHAPTER 23

Although Lynsi had remained out of the spotlight for most of her life and had only nominally participated in In-N-Out's business decisions, gradually the twenty-three-year-old heiress began making more of her presence known in the company. She was present when on December 30, 2005, In-N-Out Burger reached yet another milestone: the opening of its store number two hundred in the rural town of Temecula in Riverside County, near Camp Pendleton. On that dry winter morning, there seemed to be a sense of history coming full circle. Among those gathered at the pre-opening party was Harry Snyder's nephew Bob Meserve. He had started out picking up trash at In-N-Out Burger; some forty years later, he was the chain's director of new stores. Jack Williams, who had spent many hours horseback riding with Rich Snyder on his own Temecula ranch twenty years earlier, arrived astride his horse to celebrate. Among locals, the occasion elicited a sense of collective achievement as well as excitement. City councilman Michael Naggar showed up dressed in a shirt decorated with stars and stripes.

In fact, In-N-Out Burger had a lot to celebrate. At the time of the Temecula ribbon cutting, the chain was generating an estimated $350 million in sales annually, up a healthy 6.7 percent from the previous year. In 2005, according to the food industry consulting and research firm Technomic Inc., In-N-Out stores were raking in an estimated

$1.79 million, making the chain one of the top earners among the nation's fast-food joints and placing it second behind McDonald's, whose store sales were $1.98 million.* With plans under way to move further into new territory with the proposed Washington City, Utah, store, there were quiet rumblings that In-N-Out might be switching gears and planning a push into the southwest too, possibly even moving farther north and east.

On the top floors of University Tower in Irvine and in the second-floor executive offices in Baldwin Park, the picture was not nearly so sanguine. Several weeks before the Temecula opening, longtime In-N-Out executive and Snyder family friend Richard Boyd filed a lawsuit in Los Angeles Superior Court against the company, Lynsi Martinez, Mark Taylor, and several other executives. Boyd, a vice president at the burger chain, a board member, and cotrustee of the Snyder family trusts, charged that the company's top management was involved in a conspiracy to expel him from the company. The plot, he alleged, was actually part of a broader scheme intended to wrest control of the company from the ailing Esther Snyder in order to speed the succession of her granddaughter Lynsi. With Boyd out of the way, Lynsi would have free rein to ramp up the chain's expansion outside of its core market—or possibly, his suit suggested, even take the company public.

At the time, Lynsi was already a significant shareholder. She owned 23.59 percent of the corporation's stock independent of the trusts (Esther's share, per the agreement, was reduced to 3 percent). In two years, Lynsi would begin receiving the first installment of the family trusts, valued at $450 million. It would be another twelve years before she owned the company outright. According to the lawsuit, Lynsi was not alone in the bid to accelerate her inheritance; Boyd contended that she was colluding with Mark Taylor and various

---

* Burger King came in eighth and Wendy's, sixth.

members of In-N-Out's top executive team. Those officers, he claimed, "were fearful of losing their jobs and benefits." Their machinations, he charged, undermined not only the expressed wishes of Esther Snyder but also were in contradiction to the terms spelled out in the set of irrevocable trusts.

Not surprisingly, the explosive allegations were deeply troubling to the burger chain that had long viewed itself as a large extended and happy family. Ostensibly to quell any sense of internal strife, the family broadcast a video message to their associates on *Burger TV*; it showed Lynsi sitting with her grandmother while Taylor rallied the troops. "We're in good shape," he said. "Don't believe everything you hear."

On December 9, 2005, two days after Boyd first filed, he withdrew his suit. There was a brief period of calm as the two parties and their teams of lawyers began settlement discussions. These broke down quickly. On January 5, 2006, Boyd returned to court and re-filed his original suit. A week later, In-N-Out countersued, accusing Boyd of fraud, embezzlement, and the misuse of company funds as well as breach of contract and fiduciary duty. On January 30, In-N-Out fired Boyd from his position as vice president and removed him from the board. The legal fight consumed much of the next year and a half. Before it was over, the famously private burger chain had been mercilessly scrutinized, its future growth strategy was threatened with exposure, and a pitched internal power struggle had erupted into an ugly exchange of allegations, mudslinging, and assertions of deceit and betrayal.

At the crux of the fight seemed to be the classic contest over power and money, pitting the past against the future and longtime managers against blood descendants. The portrait that emerged in court filings placed In-N-Out at the proverbial crossroads. Virtually unchanged since 1948, would the regional chain continue its slow, steady growth or rapidly evolve into a national brand? Boyd represented the past; he was part of the founding generation of the In-N-Out dynasty, safeguarding the principles that Harry and Esther Snyder established and that their sons, Guy and Rich, maintained. Lynsi was the restless heiress, part of the third generation who wanted the riches without having to work for the rewards, while Taylor, a hapless and ambitious parvenu, lucked into the company by marrying into the weakest branch of the family tree.

At stake was the very future of In-N-Out Burger. The clash was of the sort that had fascinated social critics and observers throughout time: the fight among generations to seize control of the family fortune. Of course, it was exactly the kind of struggle that had poisoned relations and ripped apart numerous venerable family coffers. In recent times, such bitter infighting and legal wrangling had left the fortunes of the Haft family of Washington, D.C.'s estimated $500 million to $1 billion retail empire, the Chicago-based Pritzker family's $15 billion real estate holdings, and the Mondavi family's $1 billion Napa Valley winery fortune in tatters, with as much bad blood shed as ink spilled by the press chronicling their denouements.

Of course, there were two bitterly contested versions of the In-N-Out rift. Not long after the initial slew of filings, Lynsi released a statement. Boyd's lawsuit, she declared, "contains outright lies and awful inaccuracies to try and cover his errors. By far, the most upsetting is his fabrication about the relationship between me and my gramma." She took particular umbrage at his claim that she was attempting a coup against Esther, who she said should be able to "continue to serve as president as long as she desires." Boyd's attorney, Philip Heller, scoffed at the charges leveled against his client. "We sued," explained Heller, "and then they brought this ridiculous claim that was demonstrably false and vicious."

The feud began around 2003. It was then, according to court documents, that Lynsi—who had minimal involvement in company matters—began "attempting to exert control over [the chain] as well as [its] directors, officers, and employees." According to Boyd's suit, he had thwarted Lynsi's attempt to print up In-N-Out business cards that identified her as the chain's "owner." She held a minority stake in the company, as did Esther, who also happened to be the company's president, and Boyd thought the move both disrespectful and inappropriate. On several occasions before he died, Boyd claimed that Guy Snyder had told him that he didn't want his daughter to receive her inheritance before she was mature enough to handle it; he wanted her to understand the responsibility and receive an education first. However, Boyd's efforts to encourage Lynsi to "actively learn every aspect of In-N-Out operations and to attend the meetings of the In-N-Out vice presidents" in preparation for her eventual control of the company, he contended, seemed to create a further rift between the two.

It was around this time that Lynsi's religious life was said to overlap with her business affairs. The chain, previously closed only on Christmas Day and New Year's Day, was now closed on Easter Sunday as well, reportedly on Lynsi's instructions (she was also said to have asked for the antique bar to be removed from the Liberty Room in Baldwin Park). According to the filings, Lynsi held weekly prayer meetings for associates in her home that featured prerecorded sermons by Steven A. Radich, the Successful Christian Living church's Apostle. Her husband, Richard Martinez, asked In-N-Out associates to pray with him during the workday. Boyd took issue with what he considered the couple's "attempt to foist [their] religious beliefs on [their] employees." Boyd contended that Lynsi "did not believe [him] to be a man of God and worked to have him removed from the company."

By early 2004, the discord between Lynsi and Boyd seemed to have reached its apex. Boyd charged that she used Taylor to convey her wishes, instructing her brother-in-law to "fire, demote, or transfer to other departments those she believed to have slighted her." Boyd was apparently one of these. According to court documents, Mark Taylor and Roger Kotch, the chain's chief executive officer and vice president for administration and finance, approached Boyd and asked him to resign from his position as cotrustee of the Snyder family trusts. Although the two men indicated that Boyd could remain as vice president for real estate and development, they claimed to be acting on behalf of Lynsi, who wanted to remove Boyd and install Shawn Prince (the husband of her half-sister Terri) as cotrustee. Angered by the request, Boyd refused to consider the suggestion until Lynsi explained to him personally why she wanted him to resign.

On April 8, 2004, Boyd asked to meet with Lynsi, Taylor, and a complement of lawyers to discuss matters surrounding taxes owed on Guy Snyder's estate. After Guy died, his estate posed a huge tax burden that had the potential to financially wipe out the company. In order to reduce about $47 million in taxes from his estate and pay off the outstanding federal and state taxes owed, Boyd and Taylor, as cotrustees of the family trusts and co-executors of Guy's estate, along with their lawyers worked out a plan for the trusts to borrow about $60 million from the company to pay it off; the result was an agreement whereby the trust would be required to pay back about $147 million to the company, including interest, in 2015. However, Boyd claimed that Lynsi refused to attend the meeting. Taylor and Kotch explained to Boyd, according to the filings, that Lynsi had told them that if Boyd intended to be present, she would not "be in the same room with that son of a bitch."

According to the suit, a week later, on April 14, Taylor informed Boyd that Lynsi was asking whether he had reconsidered resigning as cotrustee. He was asked to take an early retirement and pressed again to resign from the trust in order to allow Shawn Prince to be installed in his place. Boyd remained defiant. Adding Prince, he countered, directly violated the terms of the trust, which did not permit the naming of successor trustees. Furthermore, Boyd asserted that Prince was "unqualified to administer a multimillion dollar trust." After Boyd flatly refused, he was told that Lynsi exclaimed, "Doesn't that green-eyed monster know what the Snyder family wants?"

From that point on, Boyd contended that he was left out of vice presidents' meetings and not consulted on a spectrum of management issues. Boyd asserted that he found out the reason from other vice presidents, who were performing a delicate balancing act; Lynsi now refused to be in the same room with him. Beginning in June, Boyd was no longer asked to take documents to Esther's home for her to approve and sign. Since Esther had been largely homebound after breaking her hip, Boyd had regularly brought corporate papers and snapshots of new store openings for Esther to review. During this time, Boyd contended, he was routinely denied information and documents that he needed to perform his duties as cotrustee. Eventually, he claimed, Taylor replaced the long-standing counsel of the Snyder family trusts without notifying Boyd.

Increasingly, Taylor was exercising his ambitions within the highest ranks of the company. During this time, Boyd asserted that Taylor was transparent in his desire to become chief executive of In-N-Out. He seemed bent on making his mark on the chain to which in reality he had only a tenuous claim, given the trusts' mandate to ensure that the company be passed on to direct blood heirs. Taylor, Boyd charged, was aggressively pushing to transform In-N-Out into a national and later international company, "without regard to In-N-Out's ability to service those markets" or "ensure In-N-Out's traditional superior quality and service."

In a company that had long made decisions by consensus, Boyd asserted that Taylor took to launching his own initiatives without consulting other vice presidents. At one point, Boyd claimed that Taylor had tried to differentiate the egalitarian uniforms of managers from associates—a move, he contended, which would have "substantially changed In-N-Out's corporate culture."

Sometime during 2004, Taylor reportedly attempted to have Esther put in a home. The effort was thwarted only after her nephew Joe Stannard (who had power of attorney over Esther's medical care) interceded. The episode did not exactly endear Mark to Esther, who Boyd alleged had become "extremely angry and hostile to Taylor."

A campaign of isolating Esther, Boyd charged, began in earnest. Taylor and other executives neglected to inform or consult her concerning business decisions and developments at the chain including price increases and company growth, while at the same time conducting management meetings without her knowledge or presence. During this time, Boyd insisted that he continued to consult and advise Esther regularly about business developments, visiting her at her home or discussing matters over the telephone.

Although she lived less than two miles from Esther, several intimates said that Lynsi saw little of her grandmother. During a meeting concerning Esther's health care, Joe Stannard urged Lynsi to visit. "She said, 'I don't know how to talk to an old lady,' " he re-called. " 'And she bugs me about college and working at In-N-Out all the time.' "

By the fall of 2005, the atmosphere within the company's executive ranks had grown perceptively tense. The company initiated an internal investigation—as a result of the inquiry, In-N-Out claimed that Boyd, with twenty years of company service under his belt, had allowed construction costs for new stores to escalate. It also maintained that he favored one contractor, Michael Anthony Companies, with uncompetitive bids. On September 16, 2005, Boyd was given a notice of non-renewal of his employment, signed by Esther Snyder. When Boyd showed Esther the notice, according to court filings, she said that she didn't recognize it. She did, however, recall that Taylor had brought a document to her house for her to sign a day earlier without explaining what it was. The incident seemed to upset Esther, who Boyd insisted said that she didn't want him to leave the company. Shortly after he left her house, Boyd claimed that Taylor telephoned him, angrily shouting, "How dare [you] upset an old lady?"

Following this episode, Boyd contended that his contact with Esther became less frequent and his ability to get in touch with her was increasingly limited.

After being marginalized for some time, Boyd asserted that by September 2005, Esther was essentially a prisoner in her own home. Guards supplied by In-N-Out were posted on the property, and only pre-approved visitors were allowed inside. Her phone calls and correspondence were screened, and at one point, Boyd alleged, Esther's telephone line was disconnected and Boyd's own telephone numbers were blocked, preventing any incoming calls to Esther from going through. (Others contended that her number was changed to prevent solicitation calls.) A grainy, blurry picture that Boyd said showed guards in position outside of Esther's Oak Tree Terrace home was submitted to the growing volume of court filings.

About six months later, in February 2006, after Esther's old friend Wolf Kahles read an article about the ongoing legal battle between Boyd and In-N-Out Burger in the *Los Angeles Times*, he became concerned and telephoned Esther from his home in Germany. The pair had talked frequently; however, when he attempted to call her private home line, he found that the number had been changed and was unlisted. "I was upset," he recalled. "They changed her telephone number and didn't tell anybody." Kahles called In-N-Out's headquarters only to be rebuffed. "I asked for her number and they said no. Then I left her a message with them and told her to call me. They never gave her any of my messages."

On November 5, 2005, In-N-Out Burger retained Grant Thornton, a large tax firm, to perform a forensic accounting analysis on Boyd; specifically, the firm was tasked with looking into allegations of inappropriate transactions. The firm was given access to Boyd's department offices and files. They interviewed Boyd's staffers and a number of the contractors and subcontractors, including Michael Madrid, owner of Michael Anthony Companies. The contents of Boyd's desk and those of the associates from his department were examined and copied. In addition, they reviewed the information stored in his computer and retrieved the laptops of Boyd's staff. The accounting firm also undertook a public record search of Boyd, his wife, son, daughter, and sister. Five days later, In-N-Out notified Boyd that it would hold a hearing on December 13 to determine whether he would be terminated as a result of "misconduct."

A month later, Lawrence A. Rosipajla, Grant Thornton's director, delivered a thick seventy-eight-page report to In-N-Out complete with copies of invoices and photographs of Boyd's second home in Arizona dated December 5, 2005. Among the findings, the report alleged that Boyd used his favored contractor, Michael Anthony Companies, to perform construction work on his personal residence but charged the cost of the work to In-N-Out. The investigation also turned up what it claimed was evidence that Boyd had allegedly embezzled company funds to build a wall costing more than $5,000 on his property in Arizona and a patio and cabana at his home in California. To cover his tracks, the report claimed that Boyd shredded documents. Boyd's attorney Philip Heller described the investigation as a charade, tantamount to character assassination. It was, he said, "nothing more than a witch hunt"; and Boyd's team soon delivered a counter investigation by Discovery Economics, refuting Grant Thornton's findings.

As the fall turned to winter and 2005 turned into 2006, the situation between Boyd and In-N-Out Burger grew nastier still. In January, the board of directors called a special meeting, the purpose of which was to discuss Boyd's termination. At the time, In-N-Out's board was composed of Boyd, Taylor, and Esther. (Boyd and Taylor were named to the board following Guy's death.) It was an odd situation to say the least. Boyd countered that he should be present and Esther included and attempted to hold the confab on January 27, later failing to get a temporary restraining order enjoining the board meeting scheduled by In-N-Out for January 30, 2006, at Esther's home. Boyd claimed he was prevented from entering the house by guards. He was not allowed to speak with Esther and was told that she didn't wish to see him.

The battle moved to the chambers of the Superior Court in downtown Los Angeles. In March, Boyd's complaints grew to include a defamation suit against the accounting firm Grant Thornton, allegedly for making "intimidating and misleading statements to In-N-Out employees." During the course of the Thornton investigation into Boyd, the firm (his attorneys charged) branded Boyd a "thief" and labeled him "unethical" during interviews with In-N-Out employees. Further, Thornton's director Rosipajla "intimidated and terrorized employees by threatening them with criminal prosecution, civil penalties, and adverse employment action if they refused to state that Boyd had engaged in criminal acts." It was hard to reconcile the behavior listed in the court documents with In-N-Out's squeaky-clean reputation.

In-N-Out extended its lawsuit to include Michael Anthony Madrid, the contractor accused of receiving preferential bids and doing work on Boyd's personal property (work that the chain's attorneys claimed was invoiced to In-N-Out). In a lengthy declaration signed and dated on October 31, 2005, Madrid explained that he had worked on over one hundred construction projects for the chain since submitting his first proposal to build the Thousand Palms store about twenty years earlier. In-N-Out, he explained, made up roughly 25 to 30 percent of his firm's business. "In-N-Out is a very important customer," he stated. "I would not do anything to jeopardize that relationship."

According to the declaration, in late 2003, Boyd told Madrid that he wanted to build a wall on his Arizona property, which happened to be twenty minutes from Laughlin, Nevada, where In-N-Out was building a store. Boyd, Madrid said, had asked whether it might be possible for Madrid's crew to work on the wall while it was working on the nearby store. According to Madrid his firm didn't create a job number for the wall because it was such a small project. Furthermore, he said, he never sent Boyd an invoice for the job, nor did he ask him to pay for it.

Then, sometime between August and September of 2005, two men appeared at the Michael Anthony Companies headquarters in Bloomington, California. The gentlemen said they were from In-N-Out and were conducting an audit. They questioned the firm's employees about its pricing process and explained that they were looking into the burger chain's projects in four cities. They probed at length about the Laughlin store, asking "pointed questions" about Boyd's wall project. After being questioned for nearly an hour, Madrid asked the men to produce their business cards and explain who they were. One identified himself as an attorney. The other man said that he was In-N-Out's controller. After another thirty minutes, Madrid became concerned about the line of questioning. "It appeared to me that this was not just about an audit, as I had been led to believe," he stated. "It really is none of their business how we use funds paid to us." Then Madrid escorted the men to the door, telling them, "I am not sure what it is you're digging for, and maybe I should have counsel to help me determine that."

A trial date was set for October 17, 2006. Before that time, a series of hearings were held to settle the score regarding Boyd's dismissal and his continued role as cotrustee. The bitter fight went on for several rounds. On March 30, Judge Mitchell Beckloff suspended Boyd as cotrustee of the family trusts, given the conflict between Boyd and Lynsi, and appointed Northern Trust Bank of California to serve as cotrustee with Taylor pending a hearing to settle the matter on a permanent basis. Round one went to In-N-Out. Then, on April 5, the burger chain suffered a setback when Judge Aurelio Munoz threw out two of its lawsuits against Boyd—breach of contract and breach of fiduciary duty—saying that they violated Boyd's free speech rights. (In-N-Out had contended that Boyd's suits revealed trade secrets and breached his confidentiality agreement.) For Boyd, the court's decision was like a moral vindication.

It was only a partial victory, however; on April 28, Boyd lost his fight to remain employed with In-N-Out. Judge Munoz ruled that Boyd could not force the company to reinstate him. "This is an action for a wrongful discharge," the judge stated, "not an action to restore a prince or a king to a throne." This time, the legal round went to In-N-Out.

As the fall trial date loomed, the ugly and recalcitrant claims spurred on for several months and filled thousands of pages of legal filings. Lynsi, Taylor, and In-N-Out filed a motion in February to have several descriptions that they called "irrelevant, sensationalized, and unfounded allegations" removed from the lawsuits. On April 28, Judge Munoz ruled to strike several of them from the legal record.

However, his decision came after what the company deemed "dirty laundry" had already aired publicly. What emerged was a family drama filled with jealousy, scheming, splintering loyalties, money, and tragedy. Boyd claimed that Lynsi did very little work at the company but expected In-N-Out employees to perform errands for her such as cleaning up after her dogs on the company's dime. Lynsi's personal life came under particular scrutiny; she was called "reckless, impetuous, irresponsible," among other personal shortcomings.

Boyd's assessment of Lynsi's business acumen and work ethic was no less damning. He complained that the girl who had all the world laid out before her if only she would lift her fingers to grab it did not take her work seriously. "She is often late for work and meetings or does not show up at all." Calling her the "Hamburger Princess," a sobriquet that most likely needled the young woman, he claimed that "despite her financial resources, Lynsi chose not to pursue a college education." Boyd also asserted that Lynsi "lacks sufficient personal maturity, experience, and skill necessary to successfully run In-N-Out." Her main qualification for running the company, he charged, was simply that she was the granddaughter of its founders.

Like a man expelling air after holding his breath for years, Boyd's descriptions spared few. Lynsi's mother, Lynda Snyder Kelbaugh (now married for the third time), was portrayed as a combination of Lady Macbeth and Cruella de Ville who had long sought to control the company, first through her son-in-law Taylor and now, as she was coming into her inheritance, through Lynsi. He charged, among other things, that "Lynda exerted her influence to push Guy Snyder to name Mark Taylor as cotrustee of the family trusts."

Boyd himself reportedly became ensnared in Lynda's wrath when he insisted several years earlier that Esther, recovering from her hip surgeries, should be removed from the hospital in Redding where she had been for nearly a year. Fearing that she would languish up north, or worse, he pushed to bring her home to Glendora, where he said she would receive better individual care and be more comfortable. A combative Lynda phoned him, he claimed, arguing that Esther should be left to mend where she was, not far from Lynda's ranch in Shingletown.

Mark Taylor appeared to be a slightly more complicated fellow. Stocky with a compact frame, his bald head gave him something of the look of a character in a 1940s gangster movie. Depending on the view one took, he was either a simpering sycophant tethered to the skirts of his mother-in-law (who fanned his ambitions for her personal gain) or was nakedly determined in his own right, short on intellect but long on cunning and street smarts. Then again, he may have just been a hardworking guy who got incredibly lucky.

Married to Traci Perkins, Taylor considered himself Guy's son-in-law and "virtually an apprentice" to Rich Snyder. Indeed, Rich was said to have respected him as a manager. In court filings, Taylor detailed his own lengthy association with the Snyder family and his rise to the top at In-N-Out from his start in 1984, to becoming a store manager, to being made manager of operations in 1998. In 1999, he was named successor cotrustee to the family trusts, and in 2001 he was named to the board of directors. As he was working his way up through the ranks to become a vice president, Taylor asserted that he had become intimately familiar with the company's culture and traditions and a variety of roles in operations as well as the family trusts.

Upon his ascent, Taylor moved into Guy Snyder's five-bedroom, six-bathroom Glendora house. Apparently, Taylor purchased the Mediterranean-style property under market price after Guy's death and set about implementing an extensive remodel. Even among the neighborhood's other outsized homes, the remodeled Taylor estate was impressive. High in the foothills of the San Gabriel Mountains, the house occupied an entire corner lot. The property, dotted by palm trees, had a large, circular brick driveway with a stone, Spanish-style fountain similar to the one in front of the San Gabriel Mission. The enormous entryway looked like the portico of an old Spanish hacienda; there were wrought iron lanterns and perfectly clipped lawns. The renovation and landscaping were so all-encompassing that they required entitlement approval from the city.

Boyd later claimed that Taylor's remodeling project cost $2 million and that he used In-N-Out vendors, suppliers, and employees to perform the work and improvements either at a reduced rate or without any compensation. In some cases, Boyd charged, Taylor billed In-N-Out for the work. At the same time, Boyd asserted that Taylor had $160,000 worth of landscaping charged to an In-N-Out construction project. When Boyd found out, he informed the company's other executives, who then pressed In-N-Out to "loan the money to Taylor" for the services. Among company executives, apparently the practice of using In-N-Out's contracters was said to be fairly commonplace.

According to Boyd's filings, Taylor was a "lackluster manager." Little more than a "rubber stamp" cotrustee when it came to dealing with the complicated matters of the family trusts, Boyd claimed that Taylor deferred or largely delegated his trust-related responsibilities to Boyd. Taylor rarely read documents or reports, Boyd claimed, and attended meetings of the cotrustees with little or no preparation. Boyd expressed grave concern over Taylor's ability to administer the trust going forward. He charged that Taylor never thoroughly read or examined the secured promissory note for $147,153,441.23 or the pledge agreement that allowed the trusts to borrow money from the company to pay off Guy's estate taxes; that in fact, he "only signed those documents after Boyd and the Trust advisors approved them." If Boyd were removed as cotrustee, he contended, "funds might not be set aside to repay the note."

Boyd insisted that Taylor's ambitions overshadowed his abilities. Furthermore, he claimed that he wasn't alone in this assessment; before he died, Guy Snyder made it clear that he didn't think Taylor should be promoted above the level of operations manager and should certainly not be made head of the company. Boyd asserted that Guy considered his former son-in-law "too ambitious and insufficiently talented to lead In-N-Out."

In his own court filings, Taylor denied Boyd's allegations. In the end, Esther would have the last word.

CHAPTER 24

It wasn't long before the battle was splashed across the pages of newspapers and magazines across the country. At In-N-Out corporate headquarters in Irvine, the inner circle fumed over what they called a "relentless publicity campaign designed to invoke the media in Boyd's war" against Lynsi. There was no shortage of coverage. "In-N-Out Lawsuit Exposes Family Rift," blared the *Los Angeles Times*. Three thousand miles away from the nearest In-N-Out, the *Washington Post* soberly weighed in with "Iconic In-N-Out Battles Executive over Firm's Direction."

The lawsuits promised to reveal not just a glimpse, but a bird's eye view into the company that had spent its entire existence largely spurning publicity. There was a great deal of public curiosity about the situation. It wasn't just that In-N-Out stood apart from the entire fast-food industry; in a time of rampant corporate malfeasance (Enron, Tyco, WorldCom), In-N-Out represented the integrity of a different era. Perhaps more than anything, In-N-Out had been around for so long that many seemed to feel that they had a personal stake in the events taking place far from public view. The burger chain occupied a particular place in the cultural imagination of Southern California, and it seemed to influence the country's perception of Southern California as well.

On January 18, 2006, when the U.S. Agency for International Development released a report depicting the security situation in Iraq as dire, amounting to a "social breakdown," concern over In-N-Out's future ranked high on the editorial pages of the *Los Angeles Times*. The chain was a "state treasure," the paper opined, "one worth preserving." Less than a month later, writer Joe Christiano dissected the corporate squabbling at the burger chain in the paper's Sunday magazine. "The whole thing seems sordid, ugly and, worst of all, familiar," he wrote. "It stinks of a monopolizing American greed that won't be satisfied with local success; it must be national, it must be global, it must be viral. Remember King Midas? Everything he touched turned to golden arches." Lamenting the possibility of a national rollout, he asked, "Will it still be ours if its bright yellow arrow points east?"

If outsiders were shocked by the allegations that threatened to destroy one of California's most beloved and iconic brands, jeopardizing its old-fashioned service, made-to-order burgers, and limited growth in favor of, well, just plain greed, the residents of Baldwin Park were absolutely horrified. Despite the myriad socioeconomic changes that had touched Baldwin Park, it still retained something of a small-town feel—especially among the "old-timers." Baldwin Park residents had long viewed the Snyder family as their own; numerous locals had worked at the chain over the years. An entire glass case in the Baldwin Park Historical Society displayed In-N-Out ephemera (wooden tokens, paper hats) alongside war veterans' uniforms, flappers' costumes, and relics of the native Gabrieleno Indians. In one way or another, at least three generations at Baldwin Park had been touched by the Snyder family.

Many seemed disturbed by the airing of the bitter feud. Others became protective and even angry, calling it a shame and hoping that the nasty business would just blow over. Those who knew both parties were particularly troubled by the accusations being lobbed. "It's not true!" later exclaimed one old family friend, who called the charges against Boyd a sham. "It's just about the money." Rich Snyder's widow, Christina, was saddened by the whole thing. "Rich Boyd is my friend," she later declared. "I've always believed in him."

For a company that had for decades kept so quiet that most of its fans had never heard the name Snyder, the public display was loathsome. For the most part, In-N-Out had dealt with internal problems more or less the way it ran its business: quietly, efficiently, and—when the occasion called for it—through negotiation and backroom dealings. Litigation was settled quietly and quickly, often with settlements and nondisclosure agreements. The $380 million company did not have a complement of full-time publicists on hand to bury the story—or even to manage their side of it.

In the midst of the public haranguing, the burger chain turned to Sitrick and Company, a crisis management and public relations outfit, for help. The Los Angeles–based firm operated under the motto: "If you don't tell your story, someone else will tell it for you," charging $695 an hour for what it liked to boast was its expertise in shaping news and public opinion. Sitrick's list of clients was an insider's guide to money, celebrity, power, and trouble. In the past, Sitrick's associates (many of whom are former journalists) have helped to deflect negative public opinion from and burnish the images of the Catholic Archdiocese of Los Angeles during the pedophile scandal of 2001 as well as embattled talk show host Rush Limbaugh when he was charged with prescription pill fraud in 2006. In a lengthy *Los Angeles* magazine profile, the firm was described as a kind of "bull in the china shop" of public opinion with a "desire not only to engage in battle, but to annihilate opponents," flooding its clients' adversaries with negative publicity.

The firm convinced In-N-Out's reluctant executives to talk to the media. On March 31, 2006, an article billed as a behind-the-scenes peek into the beloved burger chain appeared in the *Orange County Register*; a photograph of Esther sitting between Lynsi and Taylor accompanied it. Largely favorable, the piece described the chain's smiling cashiers in their spotless uniforms, its stress on high pay, and its longtime emphasis on quality control. Mark Taylor was the company's public face in the article, and during the lunchtime rush at store Number One in Baldwin Park, he demonstrated the chain's promise to deliver made-to-order burgers by ordering off of the secret menu, briefly flustering an associate. Despite the overall high-sheen gloss, the piece did not ignore the elephant in the room: the pending lawsuits. In response to Boyd's allegations, Taylor said, "It's been painful to read stories that In-N-Out will be at every corner."

There was of course another elephant in another courtroom: Esther Snyder.

She was central to the outcome of the ensuing dispute, and both sides asserted that Esther would support their claims. As she neared ninety, Esther's physical health had been on the wane for quite some time. Mentally, she displayed great lucidity, but she also suffered from memory lapses. At times, she seemed downright bewildered. Diane Brewer remembered getting a call from Esther out of the blue in July 2005. Brewer lived in Esther's hometown of Sorento, and was a member of the Free Methodist Church where Esther had worshipped as a child. Somehow, Esther heard that the church was being desanctified and shuttered. The news upset her greatly, and she tracked down Brewer. "I had to explain to her why it was closed. This is a real rural area, and there just weren't enough people to keep it going. It was quite upsetting to her. She said, 'Where am I going to go to church when I come visit?' I tried to console her, but she seemed confused. I had a difficult time getting through to her. I would get off the phone with her and she'd call back a half an hour later and ask me the same question."

As the trial date approached, everyone was acutely aware that Esther Snyder's testimony could be crucial. The question was—for whom? Esther's attorney bluntly told the *Orange County Register* that Esther would likely side with her granddaughter. Boyd, who charged that Lynsi and Taylor had over a period of years sidelined Esther, thought otherwise. Both sides produced evidence for their claims. Between January and April 2006, there was a confusing and dizzying array of documents (purportedly signed by Esther Snyder) expressing Esther's support for the actions of Taylor and her granddaughter—as well as several stating the exact opposite.

In a notarized statement dated January 12, 2006, and signed by Esther, she asserted that after having been advised by her "trusted attorney for many years" of the numerous lawsuits filed by Richard Boyd against In-N-Out, her granddaughter, and other officers of the company, she was "sad and upset that Rich Boyd would attack the company, which I co-founded and which has treated him so well, and attack my family." The statement supported Boyd's termination from the company and his elimination from the family trusts. "I no longer have any trust or confidence in Mr. Boyd and wish to see him removed as quickly as possible from any role with In-N-Out Burger, with me or my family, or with my family's trusts."

Eleven days later, on January 23, Boyd's attorneys secured a signed declaration from Esther refuting many of In-N-Out's assertions. Among them, she stated that she did not "want Rich to resign or be removed from any position." Further, she declared that she did not believe that Boyd "would ever betray me, my family, or my company." While expressing shock that criminal charges were filed against Boyd, she also maintained that she "never gave anyone permission to investigate Rich," and insisted that "Mark Taylor does not speak for me . . . I do not want Mark Taylor to run the company." Finally, she explained, "I may be in my eighties and sometimes a little forgetful, but I am very clear about what I want for the company and the statements I am making in this declaration." On a hand-scribbled note, also signed by Esther and dated January 23, 2006, Esther instructed the company to install Bob Meserve, Harry's nephew, to In-N-Out's board of directors.

During this period, more than one person in contact with Esther maintained that she was frequently given documents to sign without knowing what she was endorsing with her signature.

The series of filings, declarations, and assertions continued apace. The suit was shaping up to be quite a courtroom show, potentially creating a face-off between grandmother and granddaughter. In February, Boyd moved to have Esther give a deposition. Not surprisingly, this proposal was vigorously contested; attorneys for In-N-Out and Esther Snyder fought strenuously to prevent her from being deposed on the grounds that she was too weak, and that the prospect of having to sit through hours of potentially stressful and grueling questioning would harm her further. Boyd's lawyers offered to allow Esther to be deposed by videotape in her home. By the end of March, Boyd claimed that he had had a telephone conversation with Esther in which she indicated her willingness to be deposed. On April 20, however, Esther signed a declaration asserting the opposite.

There was no shortage of legal muscle-flexing. On April 13, James Morris, Esther's attorney, filed a motion with the court to oppose compelling his client's deposition. He submitted letters and signed declarations outlining the failing health of the In-N-Out matriarch to the Los Angeles Superior Court from two of her physicians, Dr. James P. Larsen, specializing in internal and geriatric medicine, and cardiologist Dr. Kenneth R. Jutzy. Esther's weakened physical state, her doctors asserted, not only precluded her participation in a legal deposition but would cause her more pain and possibly even hasten her death.

A hearing was scheduled for April 26 to determine whether Esther could be deposed. As the date approached, Boyd's defense appeared to gather some critical evidence that bolstered his position. Esther's deposition could be crucial. On April 25, Esther's niece Alice Meserve Manas signed a court declaration describing Esther as anything but happy at Boyd's ouster and greatly agitated by what she was hearing about the activities going on at the corporate level. During a recent visit with her aunt, Manas stated that Esther "expressed to me her dismay that the company was not being run by 'family' and actually expressed her anger that Mark Taylor was involved at all, because he is not a member of the Snyder family." During a lengthy phone conversation between the two on April 20, 2006, Manas said that Esther told her that she had "no memory of firing Richard Boyd for cause and indicated that she would never do that—that Richard Boyd was the one she wanted to be involved with the business and the trusts." After Manas expressed her thoughts that Taylor seemed to be running the company by himself, Esther became upset and told her niece that Mark Taylor was not the "big cheese, he just thinks he's the big cheese, but he's not the big cheese."

According to Manas, the two women discussed Esther's upcoming deposition. Although she didn't take to the idea of being videotaped, Esther indicated her willingness to be deposed. And she quashed any notion that she was too ill to do so. "She has trouble getting around," Manas explained, "but she can certainly talk." Manas herself seemed perplexed when she heard about her aunt's signed declaration on April 20, which stated that Esther did not want to submit to a deposition. That, said Manas, was "in direct conflict to what she stated to me." Manas herself seemed agitated. "I am uncertain as to exactly why her voice is being silenced," she exclaimed. "Isn't she entitled to at least express her thoughts, whatever they may be, whatever she can remember?"

The shock seemed to reverberate in several corners. More than one person indicated that inside of the company, it was Richard Boyd whom Esther trusted implicitly. Indeed, in addition to being named to the board and as a cotrustee of the family trusts, Boyd claimed in his court filings that in 2004 he had been named as the coexecutor of Esther's will and as a coconservator in the event that she needed one appointed.

At the eleventh hour, Boyd's legal team produced a potentially damning piece of evidence: a transcript of a thirty-four-minute recorded telephone conversation between Esther Snyder and Elisa Boyd, Richard's wife, recorded after Esther telephoned the Boyds' West Covina home. The call came at about 5:00 p.m. on April 25, 2006. It happened to be one day before a judge was to decide on whether to compel Esther to be deposed.

Throughout the conversation, Elisa Boyd pressed Esther as to her understanding of what exactly was going on between her husband, Lynsi, Mark, and In-N-Out Burger. Esther seemed startled by the accusations against Boyd. "I can't do without Rich. . . . I can't do without Rich Boyd, can I?" she says at one point. When Elisa informed her that guards refused to let Boyd enter her house, Esther reacted with equal surprise. "They don't tell me when there's a board meeting," she exclaimed. And Esther made it clear that she didn't want Mark Taylor running the company. Taylor, she said, "wishes I was dead."

If anything, the transcript appeared to cast doubt on many of the declarations and statements that she had previously made against Boyd.

ELISA BOYD: But did you want to fire Richard, Esther?*

ESTHER SNYDER: Oh no. I want to keep him.

ELISA BOYD: You want to keep him, don't you?

ESTHER SNYDER: He's doing what good friends ought to do. Without him. . . . He defends me.

. . .

ELISA BOYD: Mark Taylor and Lynsi are using your lawyer to get signatures from you.

ESTHER SNYDER: Oh.

ELISA BOYD: Because I think every time your lawyer comes to your house for you to sign something, he tells you it's about something else and you don't read it.

ESTHER SNYDER: Umm hmm.

ELISA BOYD: And because you trust him, you sign whatever it is, and so it's always something that they are trying to do against Richard.

ESTHER SNYDER: Mmm hmm.

. . .

ELISA BOYD: And so we're trying to prove that it's not true. That you didn't fire him, that you didn't do any of that stuff.

ESTHER SNYDER: No, I didn't want him to leave.

ELISA BOYD: No. He didn't want to leave either. I mean, he didn't want to leave you. He didn't want, you know, he wanted to do what he was supposed to do, but now with all these lawyers. You know Lynsi's got seven lawyers working against Richard.

ESTHER SNYDER: Oh mercy.

. . .

ESTHER SNYDER: I, uh. I never see Lynsi.

ELISA BOYD: Oh, you never see her?

ESTHER SNYDER: Not unless I'm at a meeting or something. . . . She doesn't come by the house.

ELISA BOYD: That's too bad, because you know what, you are

---

* This is an excerpt of the entire transcript.

the one that made In and Out. If it wasn't for you, nobody
would be there doing anything.

ESTHER SNYDER: That's right.

ELISA BOYD: And Lynsi wouldn't be getting anything if it wasn't
for you.

ESTHER SNYDER: Mm hmm.

ELISA BOYD: And, um. So anyway, they're taking over the
company. They're running it right now, doing all kinds of
crazy things. But, um, I don't know, you know Richard is still
fighting—we're still gonna fight in court, and hopefully, you
know, the truth will come out. But that's the reason why we
have to record your conversations that we have with you, because they say the opposite of what you tell us. They say you
don't want to see Richard. You don't trust him.

ESTHER SNYDER: No. I want to see him.

ELISA BOYD: And, you know, they're saying that he stole
money, that he stole money from In and Out.

ESTHER SNYDER: Oh boy.

ELISA BOYD: And you don't, you know that he would never do
that Esther.

ESTHER SNYDER: No.

ELISA BOYD: He'd never steal, he wouldn't steal anything from
you.

ESTHER SNYDER: Anyway, I think they're taking all my friends
away from me.

. . .

ESTHER SNYDER: I hope that the good Lord lets me live a while
'til I get everything straightened out.

On April 26, a Los Angeles Superior Court judge ruled that Esther could be deposed in two-hour blocks over three days.

By several accounts, as the fight raged on, Esther's health deterio-
rated further. Although she was living in Arizona, Meredith Stites

visited with her grandmother once a month. Sometimes the two just sat and watched movies; during other visits they did nothing but talk. "She really liked having the company," Meredith recalled. "She talked about the office—she missed going there. She said the associates were so kind." During their time together, Esther usually asked Meredith to stay. "She told me I could stay out late and she wouldn't tell my mother. Here I was in my late twenties, living on my own. It was really sweet."

In early May, Wolf Kahles had flown to California for one of his regular visits. While there, he decided to drive directly to Esther's Glendora house, since his efforts to telephone her had all proved fruitless. When he arrived, Kahles was dismayed by what he saw. Unable to move on her own, Esther was lying on a bed in her living room with the television on. He found her weak, "isolated," and generally unaware of the events taking place around her. When Kahles told her that he had attempted to call her several times, "She didn't know that they changed her telephone," he said. Furthermore, she complained, "Now I know why I wasn't getting any calls."

When Kahles brought up the topic of the legal battle between Boyd, her granddaughter, and the company, "she seemed confused and upset," he later recalled. "She didn't know what was going on in the company. I talked to her about how they were suing Boyd and how they said that he had stolen money. She said, 'He should have come and talked to me.' She wasn't aware that he was kicked out."

Kahles left Esther's house deeply saddened, upset to find this "beautiful woman" who took such pride in her company left virtually alone in her large home. "She told me that the only joy she had was watching old movies on television," he recalled ruefully. "She called it her window to the outside." Still in possession of her chronic optimism, Esther "thanked the Lord that she still had her eyesight." As he left, Kahles said, Esther told him, "My door is always open."

CHAPTER 25

The legal endgame came to an abrupt halt with a terse press release on May 10, 2006, approximately two weeks after the transcript of the recorded telephone conversation between Elisa Boyd and Esther Snyder was entered into the record of the Los Angeles Superior Court. The five law firms representing all sides of the dispute announced that a settlement had been reached, effectively scuttling the courtroom showdown slated for October. Their statement declared, "The civil action and the probate actions have been fully and amicably resolved as to all parties to those proceedings, including Lynsi Martinez and Mark Taylor." The details of the agreement, however, remained under wraps, and neither side would comment on the deal except to acknowledge that Boyd would have "no further role with In-N-Out or the Snyder family trusts." And with that, the window into what had long been considered the most secretive company in the fast-food industry was closed once again.

Still, speculation as to the outcome was rife. Did Boyd walk away empty-handed, or with a sizable cash settlement? Did In-N-Out capitulate or prevail? The only thing certain was that throughout his numerous filings, Boyd had aired a litany of uncomfortable allegations regarding Lynsi, her mother, her brother-in-law, and the company's internal machinations. The prospect of a jury trial had promised that more would follow. It was likely that the company, which had long guarded its privacy and denied the allegations, did not relish the notion of a further public thrashing. Besides, in the weeks leading up to the settlement, Boyd's defense seemed to marshal some significant wins; a close reading of the court documents suggested that the case against Boyd might quickly unravel. It was possible that after months of legal wrangling under harsh public scrutiny, Lynsi and Taylor reconsidered the wisdom of having their dispute dredged out further. Then again, Lynsi may have had another reason to put an end to the inflammatory discord—she was pregnant.

By most accounts, Esther spent the months following the settlement largely confined to her Glendora home, surrounded by her memories, a collection of old movies that she enjoyed, and the many photographs of her family—as well as a small team of full-time caretakers. Esther's health continued to deteriorate. As she had been largely removed from the legal battles of the previous year, it was unclear whether she had an accurate understanding of how the skirmish eventually played out. Esther Snyder was still In-N-Out Burger's president, but control of the burger chain was shifting to the third generation.

Meanwhile, in a Scottsdale, Arizona, hospital room 370 miles away, another fast-food legend was drawing to an end. On Tuesday, June 6, Esther's dear friend Margaret Karcher, the wife of Carl's Jr. co-founder Carl Karcher, died after a lengthy battle with liver cancer. She was ninety-one years old. Within days of receiving the news, Esther sent Margaret's husband a one-page condolence letter on In-N-Out letterhead. According to the Karchers' daughter Barbara Wall, Esther expressed how much the couple had meant to her as friends and her gratitude for the fact that they were always there for her, especially after Harry died, and later, following the deaths of her sons. She also expressed one of Karcher's own sentiments. "She wrote that she saw Dad not as a competitor, but as a colleague," recalled Wall. "He was happy that she took the time to write the note." (On January 11, 2008, a year and half later, Karcher died just four days shy of his own ninety-first birthday, a casualty of Parkinson's disease.)

The summer of 2006 was a difficult one on several fronts. A record heat wave enveloped California. As casualties mounted in Iraq and war raged between Israel and Hezbollah in Lebanon, the Southland's attention was largely fixed on the unrelenting heat that had for most of July pushed thermometers well past 100 degrees, creating a blanket of smog that traveled from California to Maine. The air was so stiflingly toxic that more than a hundred deaths resulted. The prolonged scorcher strained the state's power grid. Authorities took to the airwaves, issuing warnings to residents and asking them to limit their physical activities, drink plenty of fluids, stay indoors, and, when possible, not drive. While there was a host of many other pressing issues of the day, it was the spectacle of the summer swelter that hung in the air like the brown haze.

On Friday, August 4, 2006, the record-breaking temperatures had just begun to recede when another attention-grabbing story captivated the region, casting a further pall over the Southland. In-N-Out Burger announced that Esther Snyder had died; she was eighty-six. Flags at the company's headquarters and at its more than two hundred shops dotting California, Arizona, and Nevada flew at half-staff.

The turbulence kicked up in the wake of the litigation between In-N-Out and Richard Boyd had barely quieted down in the three months since the two parties agreed to an undisclosed settlement. News of Esther's death pushed the reticent chain once again into the public spotlight. The company released a heartfelt public statement that began, "All of us at In-N-Out will mourn the loss of Mrs. Snyder, who provided strength and inspiration for the company and its associates." Lynsi, who a few months earlier had spent a weekend in San Diego with Esther, also paid tribute to her grandmother. "She had so many great qualities; she was sweet, warm, loving, strong, and smart," she said in a rare public statement of her own. "She had the biggest heart that stretched over miles and miles."

Esther's funeral was a private affair; there was no memorial service. A security cordon ensured that only the three hundred invited mourners gained entrance to the La Verne Church of the Brethren (about nine miles from Esther Snyder's Glendora home). The Calvary Chapel's pastor, Chuck Smith Sr., presided over the service. Esther had touched many, and she was remembered as a gentle and kind woman who through sheer determination of spirit not only helped to build a small, iconic empire but also maintained it without sacrificing her principles. Calling Esther "mom," Lynda Snyder Kelbaugh and her daughters Terri and Traci gave eulogies. Lynsi nervously took the podium, telling the mourners that she loved and missed her grandmother. None of the members of the extended Snyder or Johnson families spoke. A video montage of Esther's life was played, showing her with Harry and her sons. Following the service, the assembled went into the church's courtyard. Fittingly, an In-N-Out trailer was set up to serve hamburgers.

Not long thereafter, while going through Esther's house and readying it for sale, Joe Stannard came upon his aunt's briefcase, the one that she carried every day. Tucked inside the well-worn bag, among the various business papers, Stannard found a cache of love letters that Harry had written to his wife nearly sixty years earlier.

Word of Esther's passing quickly spread across the country. By the time that she died, In-N-Out had become an iconic American cultural institution, massively beloved: a legend. In-N-Out's loyal and besotted fans, most of whom had never even seen a picture of Esther, reacted as if they had lost a member of their own family. The saddened minions went into a kind of electronic mourning. "God bless Esther Snyder," posted one on the Internet. "Still the #1 fastfood hamburger ever created, God Bless the Snyder family . . . they didn't sell out to big corporations." proclaimed another. "You folks changed many a life with your burgers." A columnist at the *Orange County Weekly* wrote that Esther Snyder "is known in my house as the Greatest Woman Who Ever Lived." And on and on it went.

The Baldwin Park city council suspended its previously scheduled meeting twelve days after Esther's death in honor of her generosity and many contributions to the place that In-N-Out had called home from its very inception. The mayor of Baldwin Park, Manual Lozano, described Esther's passing as "a great loss to the city."

Within days, In-N-Out's founding matriarch had been memorialized in the *New York Times*, the *Washington Post*, the *Wall Street Journal*, and the *Los Angeles Times*. Even CNN and NPR mentioned her passing in their national broadcasts. Remarkably, when her husband, Harry Snyder, had died thirty-one years earlier, there was no national trumpet blast. He received a brief mention under the "Death Notices" page in his hometown paper, the *San Gabriel Valley Tribune*. It was a standard announcement that ran to only sixty-three words. In the spare, efficient manner of newspaper obituary pages everywhere, the piece listed his other surviving siblings and the time and place where funeral services were to take place.

Although Esther Snyder did not possess the kind of public face of such fast-food titans as McDonald's Ray Kroc, Wendy's Dave Thomas, or Kentucky Fried Chicken's Colonel Harlan Sanders, all the same she was every inch the commanding figure that her better-known counterparts were. In fact, she may have been one of the least known and yet most beloved figures in the industry. Esther preferred to take a back seat—at least publicly—first to her husband, Harry, and later to their sons, Guy and Rich, and yet this gentle and modest woman was in many immeasurable ways the quiet, incandescent power behind the regional burger empire. She never prevaricated; Esther stressed quality, trust in the Lord, and the importance of making sure that her associates were happy. Esther preferred to let the company she had founded with her husband six decades earlier speak for her.

For the third time in thirteen years, fans and investment-hungry suitors began asking what would happen next. Although the legal brawl was settled, with Esther gone, there seemed to be an opening for change—a generational shift was already under way. The temptation to grow from small and regional to large and national was hard to resist.

Predictably, it didn't take long for the phone in Irvine to begin ringing with a host of private equity firms and other investors on the other
end. Ten days after Esther died, Ron Paul, the chief executive of restaurant industry consultancy and research outfit Technomic Inc., did the
math for the *Orange County Business Journal*: "If the chain was worth
$300 million to $400 million, it could sell roughly $100 million to $150
million to a private equity company with the family keeping control
of the rest," he said. Already an IPO favorite, the burger chain seemed
to catapult to the top of the heap after Esther died. Not long after, Ben
Holmes (the head of Morningnotes.com, an IPO research firm) told
CNN, "It's a total cult restaurant. People would chase that one."

And yet, despite consumer fervor and clamor among investors,
such moves don't necessarily mean a happy ending. There are numerous examples of great little companies imploding on their own success. Krispy Kreme, the beloved southeastern donut chain, once had
customers waiting in the cold dawn hours before its store openings,
too. It fell from grace after its April 2000 IPO and the far-flung and
rapid expansion (including selling its donuts in supermarket chains)
that followed. The debate over exploiting a brand while keeping its
mojo was kicked into high gear in February 2007, when Starbucks
chairman Howard Schultz released a memo admitting that the coffee giant—with more than thirteen thousand outlets in thirty-nine
countries—"no longer ha[d] the soul of the past." Starbucks, he bemoaned, had come to reflect a "chain of stores vs. the warm feeling
of a neighborhood store."

In-N-Out had never lost its way. Already weary from the legal
wrangling of the past year, In-N-Out's faithful had been wondering
whether a similar fate might befall In-N-Out. "Expansion is fine,"
Thomas Keller weighed in, "but I'd probably err on the side of being
territorial. It would be nice to keep this mysterious. It's a great company. It shows what can be done when you focus on being the best."
Sitting in the Barstow In-N-Out on their way between Las Vegas and
their home in Bullhead City, Arizona, Esther Mata and her husband,
Edward, felt similarly. "There's nothing better than In-N-Out," said
Esther. "We've been coming to one for twenty years." Looking at the

snaking line inching its way toward the counter, separated by an amusement park divider, Edward chimed in, "Let me put it this way. It is a chance to remember where we came from in the 1940s, even before there was a McDonald's. We used to go to the drive-in. We'd wait here, and the waitress came on skates. We ate in our car." Even though Edward Mata was sitting on a red plastic chair inside the restaurant and there wasn't a waitress in sight or a pair of roller skates to be found, he continued, "This reminds us of what was back then, but on a small scale. We never go to McDonald's—just In-N-Out." The Matas' remarks reflected the mood in the Southland, where the hamburger had been elevated to America's culinary patrimony, that these were not idle concerns.

On Monday, August 7, 2006, five days after Esther's death, the Irvine headquarters announced that Mark Taylor had assumed the role of president of In-N-Out Burger. With Boyd out and Esther gone, it appeared that—in addition to his new position as head of the chain—Taylor was left as the only member of the company's board of directors. It was unclear when or if he would name others. Perhaps more significantly, Taylor was now the sole trustee of the Snyder family trusts (with the court-ordered Northern Trust Bank of California serving on an interim basis). In one of Richard Boyd's filings, he had described this scenario as "Guy Snyder's worst nightmare." It might have been one area where the late Snyder brothers were in total agreement. As someone close to the situation remarked, "This is never how Rich would have wanted things to turn out. He would be rolling over in his grave five thousand times."

Taylor released a statement shortly after heading up the chain in which he personally promised that "the family is absolutely committed to keeping the company private and family-operated." He assured everyone that In-N-Out would continue on the path "laid out for us by our founders," growing at a moderate pace, adding only ten to twelve new restaurants each year. He also refuted the constant murmurings that the chain would now go public.

By fall, the questions raised for months among In-N-Out loyalists seemed to have been answered, at least for the time being. The hint of corporate catastrophe was pushed back. The hoopla around In-N-Out seemed to center once again on its burgers. In the early morning hours of Thursday, September 7, Paris Hilton was arrested in Hollywood on suspicion of DUI. Hilton, who had only a year earlier filmed a racy commercial for Carl's Jr., defended herself on a local radio station soon after, saying that she may have been speeding because "I was just really hungry, and I wanted to have an In-N-Out burger."

Indeed, the little burger chain continued to make its mark across the spectrum. Known as the anti-chain with the cult-like mystique, it earned plaudits from no less than the Harvard Business School and rated a full-scale analysis from the *Cornell Hotel & Restaurant Quarterly*. Even the youth of America recognized that a meal at In-N-Out would make a big impression, apparently. In November 2006, when Brian Barry was asked what famous person he would like to have lunch with, the eight-year-old told the *San Gabriel Valley Tribune* it would be Neil Armstrong and he didn't hesitate to offer to take the astronaut to In-N-Out. "Neil and I would have cheeseburgers with onions," Barry informed the paper.

In November, as Brian Barry was hoping to take his favorite spaceman to his favorite burger place, Lynsi gave birth to twins: a boy and a girl. Soon after, she purchased Esther's Glendora home. There was speculation that she might convert it into a church.

With the birth of her children, according to the terms of the Snyder family trusts, the dynasty had been preserved. The Snyder family saga that had been marked by tragedy and heartbreak had its own seemingly happy ending. There was a new generation of direct blood heirs in line to inherit the In-N-Out empire. As usual, however, the burger chain had no comment.

On Tuesday, April 22, 2008, In-N-Out finally made it to Washington City, Utah. The new store was the chain's first move into a new state since its first Arizona opening eight years earlier and store number

215. Excitement was high, especially since word first spread in 2004 that In-N-Out was coming to Utah. Anticipation reached a fevered pitch two years later when Chadder's opened in American Fork and In-N-Out took the look-alike burger joint to court.

A number of fans drove by the spot on 832 West Telegraph while it was under construction waiting for word about just when the chain would actually open for business. It came just one day before the opening; a simple sign was put out on the spot. The line formed early. By the time of its 10:30 a.m. opening, fans had created a line that wrapped around the parking lot. Two hours later, it extended to the edge of the shopping center next door. The store's manager and divisional manager stood at the entrance, welcoming guests. Employees at the various adjacent businesses came out of their offices and stores to observe what one described as "a zoo." Wyona Frazier was one of them. As she put it: "If you haven't had an In-N-Out burger, you're missing something."

And if you looked closely, you might have seen that In-N-Out's yellow arrow appeared to be pointing east.

ACKNOWLEDGMENTS

I wish to thank the many people who aided, encouraged, and sustained me during the writing of this book. At the top, I am greatly indebted to Kit Rachlis along with Karen Wada, who first let me explore In-N-Out Burger and the Snyder family on the pages of *Los Angeles* magazine.

The narrative account of this book is based on the nearly one hundred interviews that I conducted over the course of more than two years (in person, by telephone, and by e-mail), as well as research culled from thousands of pages of legal documents, oral histories, newspaper and magazine articles, archival and other documents, and photographs, books, and home movies and video recordings. (Among the many sources cited, I would like to single out John A. Jakle and Keith A. Sculle's *Fast Food*, John F. Love's *McDonald's: Behind the Arches*, and Eric Schlosser's *Fast Food Nation*.) In many instances, I returned to several individuals repeatedly for clarification—they all exhibited tremendous kindness and patience when it came to my queries.

During the course of writing this book, I made several approaches to In-N-Out Burger requesting their participation; however, they declined to cooperate. My immense appreciation goes to those individuals named and unnamed who shared their time and perceptive memories concerning Baldwin Park,

In-N-Out Burger, and the Snyder family. In particular, I wish to thank Christina Snyder Wright, Kathy Touché, and the indispensable James Berger III.

The Baldwin Park Historical Society allowed me access to a wealth of material, including the two-volume *Heritage of Baldwin Park*, compiled lovingly by Aileen Pinheiro. Robert Benbow, president of the society, gave generously of his time and memories and made himself available to answer my questions time and time again. John Coleman, president of the Bond County Historical Society, kindly exhumed valuable books and documents that helped me piece together Esther Snyder's early years and the history of the tiny coal mining town where she was born.

The indefatigable Lois Gilman unearthed a trove of information that edified the early history of the Johnson and Snyder families. I don't know anyone with a better nose for scrupulously ferreting out the most arcane and yet significant research sources. She tracked down and helped me plow through U.S. Census Bureau documents; genealogical material; naturalization, immigration, and military records; and other public archives. Every writer should have a Lois Gilman in her corner. Bill Saporito cast a critical eye to early versions of my manuscript and offered his insightful counsel that went a long way in sharpening the final book.

The many dedicated librarians at the Los Angeles Central Public Library, as well as those at the West Covina, San Gabriel, and Baldwin Park branches of the Los Angeles County Library, were an unending source of assistance and information. They hunted down documents regarding the postwar years of their cities that were most helpful. Marilyn Williams at the Bonita High School Library uncovered the school's yearbooks when Guy and Rich Snyder were students. Morgan Yates, the archivist at the Southern California Automobile Association of America provided me with piles of information on cars, the highway system, and their role in shaping Southern California. The U.S. Navy Memorial in Washington, D.C., and Rear Admiral (ret.) Jim Miller helped to outline Esther's years in the navy in general and her duty as a WAVES in particular. Ron Paul and his team at Technomic Inc. provided me with historical data about the fast-food industry as well as estimates on In-N-Out Burger over time.

At Carl Karcher Enterprises, Anne Hallock and Peggy Burgeson arranged an interview with Carl Karcher at his Anaheim headquarters on February 27, 2007. At the time of our meeting, Karcher was ninety years old, suffering from Parkinson's, and grieving over the passing of his wife, Margaret, who had died just six months earlier. Karcher's daughter Barbara Wall accompanied her father and helped convene the interview, gleaning insight into some of his comments and history. Mr. Karcher was a fast-food legend himself and a great friend of Harry and Esther Snyder, I am grateful that he took the time to share some of his memories and thoughts, although it was obviously difficult for him. Karcher died eleven months after our interview on January 11, 2008. I feel fortunate to have met him.

A tip of the hat to the countless In-N-Out associates up and down California and Nevada who cooked up and served my off-menu preference, Grilled Cheese Animal Style, never getting it wrong and always with a smile. I also wish to thank the helpful associates on the opposite end of In-N-Out's toll-free number, who cheerily and patiently fielded my many questions even if they didn't quite understand why.

I am extremely grateful to my agent Michelle Tessler, who quickly saw the potential of this story as a book and tenaciously pursued it. At HarperCollins, Genoveva Llosa, got the ball rolling; her thoughts and suggestions remained a crucial component of the final DNA of this book. Picking up the baton was Ben Steinberg, who provided me with an abundance of enthusiasm, patience, and smarts, qualities any writer should hope for in an editor and that Ben has in spades. Rounding out the fantastic team at HarperCollins: Beth Silfin, Nancy Tan, Emily Dettmer, and Carol Holmes.

John Byrne, Martin Keohan, and particularly Nick Leiber, graciously, allowed me the flexibility, space, and time to pursue the In-N-Out story while still working at *BusinessWeek*. At the magazine, Karyn McCormack offered large doses of encouragement and Matt Vella made sure that I didn't get lost on my reporting trips with a state-of-the-art GPS system.

Emerging from the end of the light of my laptop was a host of friends. In particular, Mary and Peter Vassallo provided a refuge of friendship and much needed breaks along the way. My dear friend Sue Ruopp brought the usual good cheer and as ever lent an ear with humor. Duke Sherman, my friend and a huge In-N-Out fan, supplied me with enthusiasm, companionship, and anecdotes on the long burger highway between Los Angeles and Las Vegas and beyond. Gerry and Nessa Perman unstintingly offered me familial ballast.

Of course, before the book, there was the burger. My initial introduction to In-N-Out came courtesy of the first Double-Double fan I can remember, my late father, Leonard Perman. As a child, we hopped in his car and drove to store number eighteen in Woodland Hills, California. The drive-through was more than six miles from home, and it was always worth the trip.

NOTES

PROLOGUE

Pg. 1 *"a local newspaper had once described its"*: Christie Smythe, "Deserted Core of El Con Mall May Be Razed," *Arizona Daily Star*, September 23, 2007.

Pg. 2 *"If you actually drove by the place today,"*: Matt Dinniman, blog post, http://shiveredsky.blogspot.com/2007_04_01_archive.html.

Pg. 2 *"There are so many people excited about these stores coming to town,"*: http://www.barbaralasky.com/tucson-real-estate-blog/tucson-living/tucson-in-n-outs-coming -to-town-twice/.

Pg. 3 *"Phil Villarreal, a reporter for the* Arizona Daily Star, *recalled"*: Phil Villarreal, et al., "We Got the Urge," *Arizona Daily Star*, April 26, 2007.

Pg. 4 *"This is not something that happened overnight."*: Christie Smythe, "In-N-Out's Tucson Debut Worth the Wait, Burger Fans Say," *Arizona Daily Star*, April 24, 2007.

Pg. 4 *"At nearly 10:00 p.m., there were still over one hundred cars in the drive-through lane."*: Phil Villarreal, "We Got the Urge," *Arizona Daily Star*, April 26, 2007.

Pg. 4 *"The company temporarily opened a mobile kitchen on-site to help with the constant demand."*: Christie Smythe, "Hot Spot Still In with the In-N-Out Crowd," *Arizona Daily Star*, May 24, 2007.

Pg. 4 *"In-N-Out had already sent in a team of about forty veteran employees"*: Joe Pangburn, "In-N-Out Burger's All-Stars Set Record with Tucson Opening," *Inside Tucson Business*, June 19, 2007.

Pg. 5 *"One customer, Judi Esposito,"*: Christie Smythe, "In-N-Out's Tucson Debut Worth the Wait, Burger Fans Say," *Arizona Daily Star*, April 24, 2007.

Pg. 5 *"The Tucson opening was In-N-Out's busiest to date."*: Joe Pangburn, "In-N-Out Burger's All-Stars Set Record with Tucson Opening," *Inside Tucson Business*, June 19, 2007.

Pg. 6 *"Big news for Marana today."*: Todd Kunz, "Marana Area Businesses Booming," KVOA, November 15, 2007.

Pg. 6. *"the Marana police set up a command center near the shop"*: Ibid.

Pg. 6 *"Half a dozen people staked out places near the front doors"*: "Fewer Steer to 2nd In-N-Out Burger Opening," *Tucson Citizen*, November 15, 2007.

Pg. 6 *"When Jonathan Kaye won the tournament"*: T. J. Auclar, "18 Holes With . . ." http://www.pga.com, February 2, 2006.

Pg. 6 *" 'It was crazy,' Greg Wolf, the course's head professional, later exclaimed."*: Ibid.

Pg. 7 *"The first thing I am most familiar with is In-N-Out Burger,"*: Troy Smith press conference, as reported in Doug Lesmerises, "Smith . . . mmmm In-N-Out Burger," *Cleveland Plain Dealer*, January 4, 2007.

Pg. 7 *"Rocker Courtney Love reportedly insisted"*: Karen von Hahn, "Noticed Burger Cult," *Globe & Mail*, March 27, 2004.

Pg. 7 *"A year earlier, Julia Child"*: Stacy Perman, "Fat Burger," *Los Angeles*, February 2004.

Pg. 8 *"During the 2004 NBA finals"*: Office of the Governor of California, *Gov. Schwarzenegger and Michigan Governor Jennifer Granholm Place Friendly Wager Over NBA Finals Outcome*, June 3, 2004.

Pg. 8 *"Singer Beyoncé Knowles admitted to making"*: Reagan Alexander, Carrie Bell, Jed Dveben, et al., "Caught in the Act: The Oscar After Parties," http://www.people.com, February 27, 2007.

Pg. 8 *"When discussing his favorite places to dine in Los Angeles"*: Mark Seal, "I Would Tell You . . . But Then I'd Have to Kill You," *American Way*, August 1, 2008.

Pg. 8 *"Tom Hanks rented"*: Amy Wallace, "Average Joe Versus the Icons," *Los Angeles Times*, November 7, 1999.

Pg. 8 *"What's so hip about a hamburger chain?"*: Patrick McGeehan, "Red Carpet Leads to the Drive-Thru," *New York Times*, March 7, 2004.

CHAPTER 1

Pg. 14 *"At eight-thirty on a chilly grey morning in April 1906,"*: United States Department of Immigration, *Lists of Passengers or Manifest of Aliens Seeking Admission to the US from Foreign Contiguous Territory Port of Vancouver*, April 8, 1906, Sept 24, 1915. Note that in census bureau data, the arrival date is listed variously as 1904, 1905, and 1906; according to SS *Laurentian* historical data, likely date of departure was 1906.

Pg. 15 *"A brochure from the era proclaimed"*: The Allan Line, "Illustrated Tourists Guide to Canada and the United States circa 1880"; copy can be found at http://ist. uwaterloo.ca/~marj/genealogy/allantour.html.

Pg. 15 *"At the turn of the century when Snyder père departed Amsterdam,"*: For more on the history of Amsterdam, see J. C. H. Blom and Emiel Lamberts, *History of the Low Countries* (Oxford: Berghahn Books, 2006); Jonathan Irvine Israel, *The Dutch Republic: Its Rise, Greatness and Fall, 1477–1806* (Gloucestershire, U.K.: Clarendon Press, 1995).

Pg. 15 *"many European immigrants who were pouring"*: The Canadian government took out full-page ads in British, American, and European newspapers to encourage immigration between 1896 to 1914; copies can be found at http://www.thelastbestwest. com/last_best_west.htm as well as at the Canadian Museum of Civilization, http://www. civilization.ca/visit/indexe.aspx.

Pg. 15 *"After landing in Nova Scotia"*: U.S. Census Bureau records and *Lists of Passengers or Manifest of Aliens Seeking Admission to the US from Foreign Contiguous Territory Port of Vancouver*, April 8, 1906.

Pg. 15 *"in 1909, Hendrick married Mary"*: U.S. Census Bureau records.

Pg. 15 *"Mary's mother reputedly came from"*: Snyder family home movie; Harry Snyder, interview by Rich Snyder, circa early 1970s.

Pg. 16 *"Over the next several years, Mary followed her husband"*: Ibid.; U.S. Census Bureau records 1910 and 1930; *Manifest of Alien Passengers Seeking Admission to the US from Foreign Contiguous Territories*, September 24, 1915.

Pg. 16 *"while living in Seattle, Mary gave birth"*: U.S. Census Bureau records 1910 and 1920.

Pg. 16 *"When World War I broke out in 1914,"*: Snyder family home movie; Harry Snyder, interview by Rich Snyder, circa early 1970s.

Pg. 16 *"On September 24, 1915, when Harry fils"*: According to immigration records, Hendrick carried one hundred dollars in his pocket. *Manifest of Alien Passengers Seeking for Admission to the US from Foreign Contiguous Territories*, September 24, 1915.

Pg. 16 *"By today's standards, the house wasn't much,"*: Snyder family home movie; Harry Snyder, interview by Rich Snyder, circa early 1970s.

Pg. 16 *"A census taker at the time"*: U.S. Census Bureau data 1910.

Pg. 16 *"in reality, the couple had learned"*: Snyder family home movie; Harry Snyder, interview by Rich Snyder, circa early 1970s.

Pg. 17 *"When the Snyders settled in Seattle,"*: For more on early-twentieth-century Seattle, see Clarence Bagely, *History of Seattle from Earliest Settlement to Present Time*, vol. 2 (Chicago: The S.J. Clark Publishing Company, 1916); Norbert MacDonald, *Distant Neighbors: A Comparative History of Seattle and Vancouver* (Lincoln: University of Nebraska Press, 1987).

Pg. 17 *"He had some crazy ideas,"*: Snyder family home movie; Harry Snyder, interview by Rich Snyder, circa early 1970s.

Pg. 17 *"He didn't believe anybody should be wealthy."*: Ibid.

Pg. 17 *"Ma never had much to say,"*: Ibid.

Pg. 17 *"I don't think you could find a better painter,"*: Ibid.

Pg. 18 *"We stood and watched the animals come off,"*: Ibid.

Pg. 18 *"When he left Seattle, he owed everybody and their brother money,"*: Ibid.

Pg. 18 *"In 1922, when Harry Snyder was nine years old,"*: Ibid.

Pg. 18 *"The Snyders landed in a one-bedroom house"*: Ibid.; U.S. Census Bureau data 1930.

Pg. 18 *"During this time, the Douglas Aircraft Company"*: For more on the history of Santa Monica, see Fred E. Basten, *Santa Monica Bay: The First 100 Years: A Pictorial History of Santa Monica, Venice, Ocean Park, Pacific Palisades, Topanga and Malibu* (Los Angeles: Douglas-West Publishers, 1974); Louise B. Gabriel and the Santa Monica Historical Society Museum, *Early Santa Monica* (Charleston, S.C.: Arcadia Publishing, 2006).

Pg. 18 *"By the end of the decade, the population"*: U.S. Census Bureau data for years 1920 and 1930, published by the City of Santa Monica.

Pg. 18 *"Hendrick took a series of painting jobs"*: Snyder family home movie; Harry Snyder, interview by Rich Snyder, circa early 1970s.

Pg. 18 *"When Harry was thirteen, his father landed in jail"*: Ibid.

Pg. 19 *"In addition to living in Santa Monica,"*: Ibid.; U.S. Census Bureau data 1930.

Pg. 19 *"Founded in 1905 by tobacco magnate Abbot Kinney"*: Jeffrey Stanton, *Venice California: Coney Island of the Pacific* (n.p.: Donahue Publishing, 1993).

Pg. 19 *"Venice also became associated with"*: Matthew Bialecki, ed., *The New Bungalow* (Salt Lake City: Gibbs Smith Publisher, 2001).

Pg. 19 *"At one point in 1930, the Snyder family lived in a single-family Craftsman"*: U.S. Census Bureau data 1930. In 1990, the house (originally built in 1912) was named to the Historic Sites Survey Report.

Pg. 19 *"In 1928, Hendrick Snyder quit his job painting"*: Snyder family home movie; Harry Snyder, interview by Rich Snyder, circa early 1970s.

Pg. 19 *"Harry always found a way to earn a buck."*: Ibid.

Pg. 19 *"Among his many jobs, Harry worked as a paperboy."*: Ibid.

Pg. 20 *"It was not very much money,"*: Ibid.

Pg. 20 *"I went through high school with a C-average"*: Ibid.

Pg. 20 *"I couldn't afford to go,"*: Ibid.

Pg. 20 *"Having returned to Seattle at one point, he worked"*: U.S. World War II Army Enlistment Records, 1938–1946.

Pg. 20 *"And at age twenty-nine, Harry Snyder was drafted."*: Ibid.; Snyder family home movie; Harry Snyder, interview by Rich Snyder, circa early 1970s.

Pg. 21 *"Harry was quickly sent to basic training"*: Snyder family home movie; Harry Snyder, interview by Rich Snyder, circa early 1970s.

Pg. 21 *"until only recently been used as a crude temporary camp"*: History of the fairgrounds as an internment camp and army base can be found at the California State Military Museum and Lawson Fusao Inada, ed., *Only What We Could Carry* (Berkeley, Calif.: Heyday Books, 2000); J. Burton, ed., *Confinement and Ethnicity: An Overview of World War II Japanese American Relocation Sites* (Seattle: University of Washington Press, 2002).

Pg. 21 *"A perforated eardrum"*: Snyder family home movie; Harry Snyder, interview by Rich Snyder, circa early 1970s.

Pg. 21 *"For extra cash, Harry worked on the side in the Sausalito shipyards,"*: Ibid.

Pg. 21 *"The daughter and granddaughter of coal miners,"*: U.S. Census Bureau data 1920; Bond County Historical Society, *Bond County History: A History of Bond County, Illinois* (Greenville, Ill.: Bond County Historical Society, 1979), 314–315.

Pg. 22 *"A fifth-generation Illinoisan,"*: Bond County Historical Society, *Bond County History: A History of Bond County, Illinois* (Greenville, Ill.: Bond County Historical Society, 1979), 314–315; family lineage also traced through http://www.rootsweb.ancestry.com.

Pg. 22 *"On Esther's paternal line, her great-great-great-great-great-grandfather"*: Ibid.

Pg. 22 *"Esther's maternal grandfather, Brien Molloy,"*: Ibid., pg. 315.

Pg. 22 *"The Johnson line of the family came to Sorento"*: Ibid., 314–315; U.S. Census Bureau data 1850, 1860, and 1870.

Pg. 23 *"Before his marriage, her father, Orla,"*: Per Orla J. Johnson's draft registration signed January 1917, which states he worked for the Old Ben Coal Company in West Franklin, Illinois; U.S. Census Bureau data 1920, 1930.

Pg. 23 *"Cheap bituminous or soft coal"*: John Hancock, *Village of Panama, 100 Years in the Making* (self-published, 2006), 3; U.S. Government, *Thirteenth Census of the United States, Vol. XI, Mines and Quarry, 1913* (Washington, D.C.: Printing Office), 187, Table 4.

Pg. 23 *"Shoal Creek Company's Panama Mine"*: John Hancock, *Village of Panama, 100 Years in the Making* (self-published, 2006), 3.

Pg. 23 *"Orla worked alongside his brother"*: Bond County Historical Society, *Bond County History: A History of Bond County, Illinois* (Greenville, Ill.: Bond County Historical Society, 1979), 315.

Pg. 23 *"a man named John L. Lewis"*: Sorento, 1882 to 1982 Centennial, Bond County Historical Society.

Pg. 23 *"the Panama shut down permanently"*: Directory of Coal Mines in Illinois Dept. of Natural Resources, Illinois State Geological Survey, 2000; John Hancock, *Village of Panama, 100 Years in the Making* (self-published, 2006), 21.

Pg. 24 *"Greenville yearbook inscription reads,"*: Greenville High School yearbook, *Sho-La-Hi*, 1937, pg. 16.

Pg. 25 *"In 1943, Esther joined the war effort,"*: Snyder family home movie; Esther Snyder, interview by Rich Snyder, circa early 1970s; U.S. Navy Memorial; "In Loving Memory of Esther Snyder," In-N-Out Burger corporate website, http://www.in-n-out.com/esther/.

Pg. 25 *"The WAVES were designated by special order"*: Department of the Navy, Naval Historical Center online library, "World War II Era Waves," http://www.history.navy.mil/photos/prs-tpic/females/wave-ww2.htm.

Pg. 25 *"I thought I might enjoy radio or hospital work."*: Snyder family home movie; Esther Snyder, interview by Rich Snyder, circa early 1970s.

Pg. 25 *"boot camp at Hunter College"*: Janet Butler Munch, "Making Waves in the Bronx: The Story of the U.S. Naval Training School (WR) At Hunter College," *Bronx County Historical Society Journal*, 30, no. 1 (Spring 1993).

Pg. 26 *"WAVES recruiting posters"*: Department of the Navy, Naval Historical Center online library, "Recruiting Posters for Women from World War II—The Waves," http://www.history.navy.mil/ac/posters/wwiiwomen/wavep1.htm.

Pg. 26 *"Esther Johnson spent the next three years (until November 1945) in the navy."*: U.S. Navy Memorial; "In Loving Memory of Esther Snyder," In-N-Out Burger corporate website, http://www.in-n-out.com/esther/.

Pg. 26 *"When the war ended and Esther left the navy,"*: U.S. Navy Memorial.

Pg. 27 *"Esther enrolled at Seattle Pacific University,"*: "In Loving Memory of Esther Snyder," In-N-Out Burger corporate website, http://www.in-n-out.com/esther/.

Pg. 27 *"He just came in to deliver sandwich boxes,"*: Snyder family home movie; Esther Snyder, interview by Rich Snyder, circa early 1970s.

CHAPTER 2

Pg. 28 *"Some seven thousand people lived"*: Baldwin Park Historical Society archives; Historical U.S. Census Populations of Places, Towns, and Cities in California 1850–1990; official population records were not recorded for Baldwin Park until after 1956, when it was officially incorporated into Los Angeles.

Pg. 28 *"with its lighted baseball field"*: Aileen Pinheiro, comp., *The Heritage of Baldwin Park*, vol. 1 (Dallas, Tex.: Taylor Publishing Co., 1981), 41; "Park Plunge Draws Crowd," *Baldwin Park Bulletin*, July 17, 1942.

Pg. 28 *"Running sixty trains daily"*: A period ad for the Pacific Electric reprinted in Aileen Pinheiro, comp., *The Heritage of Baldwin Park*, vol. 1 (Dallas, Tex.: Taylor Publishing Co., 1981), 31.

Pg. 29 *"With no fewer than twelve hundred performing animals,"*: Ibid.; Cecilia Rasmussen, "Cagey Entertainer's Life Was a 3-Ring Circus," *Los Angeles Times*, September 29, 2002.

Pg. 29 *"In 1927, the flamboyant Al Barnes"*: Baldwin Park Historical Society archives.

Pg. 29" *where locals had grown wary of boisterous circus employees,"*: Ibid.; Cecilia Rasmussen, "Cagey Entertainer's Life Was a 3-Ring Circus," *Los Angeles Times*, September 29, 2002.

Pg. 29 *"During a building binge, circus elephants"*: Aileen Pinheiro, comp., *The Heritage of Baldwin Park*, vol. 1 (Dallas, Tex.: Taylor Publishing Co., 1981), 227.

Pg. 29 *"the Paramount movie studio used"*: Cecilia Rasmussen, "Cagey Entertainer's Life Was a 3-Ring Circus," *Los Angeles Times*, September 29, 2002.

Pg. 29 *"The town's thrilling circus days came to an end in 1938"*: Ibid.

Pg. 30 *"upset Republican hopes, confound[ed] Mr. Gallup and surprise[d] even members of the Democratic party."*: "Baldwin Park Votes Heavily for Democrats," *Baldwin Park Tribune*, November 4, 1948.

Pg. 30 *"Between 1939 and 1945, federal spending reached"*: Kevin Starr, *California: A History* (New York: Modern Library Chronicles, 2005), 237.

Pg. 30 *"throughout the war years, some 1.6 million Americans migrated"*: Ibid.

Pg. 30 *"between 1940 and 1950, the state population swelled"*: Ibid., 238; U.S. Census Bureau statistics.

Pg. 31 *"Between July 1945 and July 1947, more than a million people"*: Kevin Starr, *California: A History* (New York: Modern Library Chronicles, 2005), 238.

Pg. 31 *"As New York is the melting pot"*: Carey McWilliams, *Southern California Country: An Island in the Land* (New York: Duell, Sloan & Pearce, 1946), 233.

Pg. 31 *"By the time the Snyders arrived, the number of inhabitants"*: Aileen Pinheiro, comp., *The Heritage of Baldwin Park*, vol. 1 (Dallas, Tex.: Taylor Publishing Co., 1981), 17.

Pg. 31 *"Since the business done by the post office"*: "All-Time High Mark Set by Post Office Receipts of $65,000 in 1948," *Baldwin Park Tribune*, December 31, 1948.

Pg. 31 *"the town now boasted eighteen different churches,"*: Classified Ads, *Baldwin Park Tribune*, October 15, 1948.

Pg. 31 *"One of the largest, Baldy View,"*: Aileen Pinheiro, comp., "Baldy View Trailer Park," *The Heritage of Baldwin Park*, vol. 1 (Dallas, Tex.: Taylor Publishing Co., 1981), 248.

Pg. 32 *"into what historian and author Mike Davis called"*: Mike Davis, *Ecology of Fear: Los Angeles and the Imagination of Disaster* (New York: Metropolitan Books, 1999), 74.

Pg. 32 *"as a result of the generous GI Bill"*: "GI Bill Turns 62 Today," Military.com, June 22, 2006, http://www.military.com/NewsContent/0,13319,102383,00.html.

Pg. 32 *"a desirable two-bedroom house with a double lot cost"*: "Baldwin Sales Total $400,000," *Los Angeles Times*, August 21, 1949; Baldwin Park Historical Society.

Pg. 32 *"generally referred to simply as 'GI homes,'"*: "Baldwin Park Tract Opened," *Los Angeles Times*, October 22, 1950.

Pg. 32 *"and were offered to veterans for 'as low as $250 down.'"*: "Three Projects Ready for '49," *Los Angeles Times*, December 12, 1948.

Pg. 32 *"a Chicago-transplant, cartoonist, and animator"*: "Disneyland Beginnings . . ." University of Southern California digital archives, http://www.usc.edu/libraries/archives/la/disneyland/.

Pg. 33 *"Early on, he took on a partner,"*: Aileen Pinheiro, comp., "In-N-Out Burgers, Inc.": Esther L. Snyder," *The Heritage of Baldwin Park*, vol. 1 (Dallas, Tex.: Taylor Publishing Co., 1981), 242.

Pg. 33 *"Already by 1940 there were over 1 million cars"*: Carey McWilliams, *Southern California Country: An Island in the Land* (New York: Duell Sloan & Pearce, 1946), 236.

Pg. 33 *"It was reported at the time that Angelinos spent more money"*: Ibid.

Pg. 34 *"And in 1970, Koulax boasted that Tommy's had raked in $1 million;"*: Burt A. Folkart, "Tommy Koulax, Builder of Hamburger Chain, Dies," *Los Angeles Times*, May 29, 1992.

CHAPTER 3

Pg. 35 *"the $1.5 billion, 3,000-unit, multinational Carl Karcher Enterprises."*: Carl Karcher Enterprises press release, 2007.

Pgs. 35–38 Karcher's bio and company history come from author interview, February 27, 2007, and Carl Karcher Enterprises, "Never Stop Dreaming A Celebration Under the Stars A Tribute to Carl and Margaret Karcher," July 17, 1991.

Pg. 36 *"I was concerned at first"*: Ibid.

Pg. 36 *"I guess we have a hot dog cart."*: Ibid.

Pg. 37 *"I always heard that if you had three or more"*: Ibid.

Pg. 39 *"Many consider the Pig Stand"*: Michael Karl Witzel, *The American Drive-In Restaurant* (St. Paul: MBI Publishing, 1994), 25, 28; John Love, *McDonald's: Behind the Arches*, rev. ed. (New York: Bantam Books, 1995), 11.

Pg. 39 *"Houston Drive-In Trade Gets Girl Show with Its Hamburgers"*: "Roadside Service," *Life*, February 26, 1940.

Pg. 40 *"On the California border with Tijuana,"*: Michael Karl Witzel, *The American Drive-In Restaurant* (St. Paul: MBI Publishing, 1994), 39.

Pg. 41 *"Dave Thomas, the founder of Wendy's,"*: Wendy's Old Fashioned Hamburgers corporate history, http://www.wendys.com.

Pg. 41 *"Even McDonald's first drive-through window"*: "McDonald's the Rise and Stall," *Businessweek*, March 2, 2003.

Pg. 41 *"In 1951, Jack in the Box introduced its own intercom"*: Jack in the Box corporate history, http://www.jackinthebox.com/drivethru/.

Pg. 41 *"Back in 1931, the Pig Stand had devised"*: Michael Karl Witzel, *The American Drive-In Restaurant* (St. Paul: MBI Publishing, 1994), 28.

Pg. 41 *"A clutch of what were called 'drive-up windows'"*: John A. Jakle and Keith A. Sculle, *Fast Food Roadside Restaurants in the Automobile Age* (Baltimore: Johns Hopkins University Press, 1999), 61–62.

Pg. 41 *"In 1954, the Stillwater, Oklahoma, shop"*: Sonic corporate history, http://www.sonicdrivein.com/history/part1.jsp.

Pg. 41 *"Soon, a host of electronic ordering devices"*: Jim Heimann, *Car Hops and Curb Service* (San Francisco: Chronicle Books, 1996), 118.

Pg. 42 *"As Esther Snyder once proudly told the* Los Angeles Times,"*: Edmund Newton, "Faithful Customers Have No Beef with In-N-Out Burger," *Los Angeles Times*, September 16, 1990.

Pg. 42 *"In fact, the drive-through became so pervasive"*: John A. Jakle and Keith A. Sculle, *Fast Food Roadside Restaurants in the Automobile Age* (Baltimore: Johns Hopkins University Press, 1999), 61–62.

Pg. 42 *"Anything he decided to do usually turned out well,"*: Myrna Oliver, "Esther Snyder, 86; Co-Founded the In-N-Out Burger Chain," *Los Angeles Times*, August 6, 2006.

Pg. 42 *"On In-N-Out's first day of business,"*: Aileen Pinheiro, comp., "In-N-Out Burgers, Inc.: Esther L. Snyder," *The Heritage of Baldwin Park*, vol. 1 (Dallas, Tex.: Taylor Publishing Co., 1981), 242.

Pg. 43 *"In recalling those early weeks, Esther Snyder once said."*: Ibid.

Pg. 43 *"Our early support came from the kind people of Baldwin Park,"*: Ibid.

Pg. 43 *"local produce growers who drove at night"*: Ibid.

Pg. 44 *"The system was based on three simple words:"*: Ibid.

Pg. 45 *"Mr. Snyder stressed quality from the first day he opened for business,"*: Ibid.

Pgs. 46–47 *"a pair of brothers—Richard and Maurice (Dick and Mac) McDonald"*: For the history of McDonald's, I relied on John Love, *McDonald's: Behind the Arches*, rev. ed. (New York: Bantam Books, 1995), 14; Ray Kroc with Robert Anderson, *Grinding It Out: The Making of McDonald's* (Chicago: H. Regnery, 1977); David Halberstam, *The Fifties* (New York: Villard Books, 1993); McDonald's corporate history.

Pg. 46 *"Sometimes I like to play a hunch,"*: David Halberstam, *The Fifties* (New York: Villard Books, 1993), 156.

Pg. 47 *"Our whole concept was based on speed,"*: John Love, *McDonald's: Behind the Arches*, rev. ed. (New York: Bantam Books, 1995), 14.

CHAPTER 4

Pg. 51 *"Pete & Jake's Hot Rod Parts in Temple City"*: Pete Chapouris, So-Cal Speed Shop history, http://www.so-calspeedshop.com/petec.html.

Pg. 51 *"We'd start at the In-N-Out on Valley,"*: Ibid.

Pg. 52 *"During their brief marriage"*: Edmund Newton, "Faithful Customers Have No Beef with In-N-Out Burger," *Los Angeles Times*, September 16, 1990.

Pg. 53 *"When he took a break, he'd go in the tiny room"*: Comments made by Don Miller during his eulogy at memorial service for Rich Snyder, Phil West, and Jack Sims, held at the Calvary Chapel in Costa Mesa, December 23, 1993 (videotape).

Pg. 53 *"sitting on the sofa watching TV"*: Edmund Newton, "Faithful Customers Have No Beef with In-N-Out Burger," *Los Angeles Times*, September 16, 1990.

Pg. 53 *"California's minimum wage was sixty-five cents an hour,"*: California Department of Industrial Relations Industrial Welfare Commission, department of finance, economic research unit.

Pg. 53 *"They take your orders and make you food. They're so important,"*: Greg Johnson, "More than Fare," *Los Angeles Times*, August 15, 1997.

Pg. 54 *"One of them was Chuck Papez."*: Greg Hernandez, "Family Owned In-N-Out at a Crossroads," *Los Angeles Times*, July 2, 2000.

Pg. 55 *"During In-N-Out's early days, the Snyders doled"*: Ibid.

Pg. 55 *"Every Christmas, Harry walked into the Baldwin Park branch of the Bank of America"*: Ibid.

CHAPTER 5

Pg. 60 *"The Pacific Electric train,"*: Cecil Adams, "Did General Motors Destroy the LA Mass Transit System?" *Straight Dope*, January 19, 1986.

Pg. 60 *"Indeed, on October 14, 1950, at 11:23 p.m., the Pacific Electric,"*: Aileen Pinheiro, comp., *The Heritage of Baldwin Park*, vol. 1 (Dallas, Tex.: Taylor Publishing Co., 1981), 45.

Pg. 60 *"The demise of the fabled Red Line"*: Bradford Snell, "The Street Car Conspiracy How General Motors Deliberately Destroyed Public Transit," *The New Electric Railway Journal*, (Autumn 1995); Cecil Adams, "Did General Motors Destroy the LA Mass Transit System?" *Straight Dope*, January 19, 1986. For more information, refer to "Taken for a Ride," PBS (August 6, 1996).

Pg. 61 *"The scandal shot straight to the Supreme Court,"*: United States v. National City Lines Inc. 334 U.S. 573, 596 (1948).

Pg. 61 *"The American really loves nothing but his automobile"*: William Faulkner, *Intruder in the Dust* (New York: Vintage Books, 1972), 278.

Pg. 61 *"By the time President Dwight D. Eisenhower"*: Dwight D. Eisenhower presidential archives, http://www.eisenhower.archives.gov.

Pg. 61 *"The interstate act became the largest public works project"*: U.S. Department of Transportation.

Pg. 62 *"the San Gabriel Valley, a thirty-mile span of the Interstate Route 605"*: Barry Cohon, "Banks of the San Gabriel," *California Highways and Public Works*, 43 (July–August 1964).

Pg. 62 *"territories as varied and fascinating as any"*: Ibid.

Pg. 62 *"it also left Baldy View with only forty-seven trailer home slots"*: Aileen Pinheiro, comp., *The Heritage of Baldwin Park*, vol. 1 (Dallas, Tex.: Taylor Publishing Co., 1981), 248.

Pg. 63 *"By God, they're all bought and paid for, too."* Richard Martin, "In-N-Out's Size No Measure of Its Stature," *Nation's Restaurant News*, May 7, 1984.

Pg. 63 *"After American Restaurant Magazine ran a cover story"*: John Love, *McDonald's: Behind the Arches*, rev. ed. (New York: Bantam Books, 1995), 20.

Pg. 64 *"This may be the most important 60 seconds of your life."*: Ibid.

Pg. 64 *"Within two years the brothers had haphazardly sold fifteen franchises"*: Ibid.

Pg. 64 *"The success of McDonald's spurred another"*: Taco Bell corporate history, http://www.tacobell.com; Funding Universe corporate histories, http://www.funding universe.com/company-histories/Taco-Bell-Corp-Company-History.html.

Pg. 64 *"In 1952, duly impressed with McDonald's operations"*: John A. Jakle and Keith A. Sculle, *Fast Food Roadside Restaurants in the Automobile Age* (Baltimore: Johns Hopkins University Press, 1999), 116–117.

Pg. 64 *"It was the same year that David R. Edgerton Jr.,"*: Burger King corporate history, http://www.burgerking.ca/en/1122/index.php.

Pg. 64 *"Then in 1954, a fifty-two-year-old, former paper cup salesman"*: Ray Kroc with Robert Anderson, *Grinding It Out: The Making of McDonald's* (Chicago: H. Regnery, 1977), 7.

Pg. 65 *"Son of a bitch, these guys have got something."*: "The Burger That Conquered the Country," *Time*, September 14, 1973.

Pg. 65 *"he soon convinced the McDonalds"*: Ray Kroc with Robert Anderson, *Grinding It Out: The Making of McDonald's* (Chicago: H. Regnery, 1977), 12.

Pg. 65 *"On April 15, 1955, Ray Kroc opened his own McDonald's"*: McDonald's corporate history, http://www.mcdonalds.com/corp/about/mcd_history_pg1.html.

Pg. 65 *"Over the next five years, Kroc built a chain"*: "The Burger That Conquered the Country," *Time*, September 14, 1973.

Pg. 65 *"Kroc was only earning a paltry 1.9 percent of the gross"*: Ray Kroc with Robert Anderson, *Grinding It Out: The Making of McDonald's* (Chicago: H. Regnery, 1977), 72.

Pg. 65 *"In 1961, Kroc asked the brothers"*: Ibid., 121.

Pg. 65 *"Kroc not only boasted that he would open"*: "The Burger That Conquered the Country," *Time*, September 14, 1973.

Pg. 66 *"In 1952, an elementary school dropout,"*: KFC corporate history, http://www.kfc.com/about/history.asp history.

Pg. 67 *"In 1964, Sanders (who had begun opening outlets in Canada and England) sold"*: Ibid.; Amy Garber, "Yum's Got the Whole World in Its Brands," *Nation's Restaurant News*, August 15, 2005.

CHAPTER 6

Pg. 69 *"Take the 3,548,000 babies born in 1950,"*: Sylvia F. Porter, "Babies Equal Boom," *New York Post*, May 4, 1951.

Pg. 73 *"While Harry's own father, Hendrick, was a tough disciplinarian"*: Snyder family home movie; Harry Snyder, interview by Rich Snyder, circa early 1970s.

Pg. 73 *"At the academy, the students were given full dress uniforms"*: William J. P. Grace, "My Brush with History," *American Heritage*, November 1996.

Pg. 76 *"On October 14, 1947, Chuck Yeager broke the sound barrier"*: Gerald Silk, *Automobile and Car Culture* (New York: Harry N. Abrams Inc., n.d.), 182.

Pg. 77 *"A year later, Alex Xydias, a B-17 engineer"*: Ibid., 183.

Pg. 77 *"In 1949, Muroc was officially converted into Edwards Air Force Base"*: Ibid.; History of Edwards Air Force Base, http://www.edwardscareers.com/about.asp.

Pg. 77 *"Clare MacKichan, one of the design engineers"*: Michael Lamm, *Chevrolet 1955: Creating the Original* (Stockton, Calif.: Lamm-Morada Inc, 1991), cited in David Halberstam, *The Fifties* (New York: Ballantine Books, 1993), 494.

Pg. 77 *"In an effort to 'create order from chaos'"*: NHRA History: Drag Racing's Fast Start, http://www.nhra.com/content/about.asp?articleid=3263&zoneid=101.

Pg. 78 *"The sport had so transcended its humble beginnings"*: "Hot Rod Fever," *Life*, April 29, 1957.

Pg. 78 *"By the early 1960s, the NHRA had over"*: National Hot Rod Association.

Pg. 79 *"The Dale was so frequently packed and noisy"*: Peyton Canary, "Car Racing Upsets Neighboring Cities," *Los Angeles Times*, May 12, 1966.

Pg. 80 *"by 1968 there were one thousand McDonald's across the country"*: McDonald's corporate history, http://www.mcdonalds.com/corp/about/mcd_history_pg1/mcd_history_pg5.html.

Pg. 82 *"The '65 Chevelle he built for Geno Redd"*: John Jodauga, "Where Are They Now," *National Dragster*, February 2, 1996.

CHAPTER 7

Pg. 84 *"Bell Labs began to create artificial intelligence"*: Eisenhower archives, "The Eisenhower Presidential Era," http://www.dwightdeisenhower.com/president.html.

Pg. 84 *"Eisenhower administration came up with its 'Atoms for Peace' campaign,"*: President Eisenhower introduced the idea publicly in his "Atoms for Peace" speech given before the General Assembly of the United Nations on peaceful uses of atomic energy on December 8, 1953; for more, read Ira Chernus, *Eisenhower's Atoms For Peace* (College Station: Texas A&M University Press, 2002).

Pgs. 84–85 *"It was none other than Walt Disney"*: Robert Schock, Eileen S. Vergino, Neil Joeck, and Ronald F. Lehman, "Atoms for Peace After 50 Years," *Issues in Science and Technology* (Spring 2004).

Pg. 85 *"one of the most influential men alive"*: "Father Goose," *Time*, December 27, 1954.

Pg. 85 *"And in 1957, Disney (in conjunction with the U.S. Navy"*: Mark Langer, "Disney's Atomic Fleet," *Animation World*, 1998.

Pg. 85 *"Automatic electric washers and dryers,"*: Caroline Hellman, "The Other American Kitchen: Alternative Domesticity in 1950s Design, Politics, and Fiction," *Journal of American Popular Culture* (Fall 2004), http://www.americanpopularculture.com/journal/articles/fall_2004/hellman.htm.

Pg. 85 *"in 1953 of the frozen 'TV Dinner'"*: James Trager, *The Food Chronology* (New York: Henry Holt & Company, 1995), 543.

Pg. 86 *"During the 1950s, when Americans snapped up"*: According to the American Frozen Food Institute, as cited in "More than Frozen Pizza," University of California Nutrition, Family and Consumer Science, Cooperative Extension, March 2, 2000, http://cekern.ucdavis.edu/Custom_Program804/More_Than_Frozen_Pizza.htm.

Pg. 86 *"The Seeman Brothers of New York"*: James Trager, *The Food Chronology* (New York: Henry Holt & Company, 1995), 543.

Pg. 86 *"as the prices for cocoa spiked, Robert Welch,"*: Ibid., 545.

Pg. 86 *"In 1959, General Foods Corporation"*: "Just Heat & Serve," *Time*, December 7, 1959.

Pg. 86 *"Kroc hired engineers and technicians"*: John Love, *McDonald's: Behind the Arches*, rev. ed. (New York: Bantam Books, 1995), 137–139.

Pg. 87 *"As early as 1931, White Castle,"*: White Castle corporate history, http://www.whitecastle.com/_pages/timeline_40s.asp.

Pg. 87 *"In 1968 there were about one thousand McDonald's restaurants"*: McDonald's Corporation 1968 financial statement.

Pg. 87 *"To supply them all, the company was using 175 different meat suppliers."*: John Love, *McDonald's: Behind the Arches*, rev. ed. (New York: Bantam Books, 1995), 333.

Pg. 87 *"After twelve years of insisting on fresh beef,"*: Ibid., 334.

Pg. 87 *"Equity Meat Company had proved that it could standardize"*: Keystone (formerly Equity Meat Company) corporate history, http://www.keystonefoods.com/history.html.

Pg. 87 *"Equity (later renamed Keystone Foods) became"*: Ibid.

Pg. 87 *"John Richard 'J. R.' Simplot, the country's largest supplier of fresh potatoes,"*: J. R. Simplot Company corporate history, http://www.simplot.com/company/origins_founder.cfm.

Pg. 87 *"I told him frozen fries would allow him"*: John Love, *McDonald's: Behind the Arches*, rev. ed. (New York: Bantam Books, 1995), 330.

Pg. 88 *"J. R. Simplot went on to build an estimated $3.6 billion empire"*: Company corporate history and "The Forbes 400," *Forbes*, September 20, 2007.

Pg. 88 *"In 1946, the U.S. Department of Agriculture required"*: USDA Fact Sheet "Focus on Beef"; John Love, *McDonald's: Behind the Arches*, rev. ed. (New York: Bantam Books, 1995), 129. (Today's USDA standards: both hamburger and ground beef can have seasonings, but no water, phosphates, extenders, or binders may be added.)

Pg. 88 *"You would negotiate a price with the drive-in"*: John Love, *McDonald's: Behind the Arches*, rev. ed. (New York: Bantam Books, 1995), 129.

Pg. 88 *"Schlosser listed forty-seven chemical ingredients"*: Eric Schlosser, *Fast Food Nation* (New York: Perennial, 2001), 125–126.

Pg. 89 *"'This way,' the company later proclaimed, enabled it to 'completely control the patty-making process.'"*: In-N-Out Burger corporate website, http://www.in-n-out.com/statement.asp.

Pg. 90 *"In-N-Out took the same approach with its french fries"*: In-N-Out Burger corporate statement on quality, http://www.in-n-out.com/freshness.asp.

Pg. 90 *"Frequently, the potatoes were picked in the morning"*: Edmund Newton, "Faithful Customers Have No Beef with In-N-Out Burger," *Los Angeles Times*, September 16, 1990.

Pg. 90 *"In an industry that was substituting chemically processed,"*: In-N-Out Burger corporate statement on quality, http://www.in-n-out.com/freshness.asp.

Pg. 90 *"TV ownership had grown exponentially,"*: According to U.S. Labor Statistics as cited in Lizabeth Cohen, *A Consumer's Republic: The Politics of Mass Consumption in Postwar America* (New York: Knopf, 2003), 19.

Pg. 91 *"McDonald's launched its first national commercial in 1967."*: "Thoroughly Modern Marketing," *Nation's Restaurant News*, April 11, 2005.

Pg. 91 *"Jack in the Box, founded in 1951,"*: Jack In the Box corporate history, http://www.jackinthebox.com; Rodney Allen Rippey's homepage at http://www.rodneyallenrippey.net/.

Pg. 91n *"In 1973, television producers Sid and Marty Krofft"*: Cecil Adams, "Was McDonaldland Plagiarized from the old HR Pufnstuf Kids' Show?" *Straight Dope*, August 27, 1999, http://www.straightdope.com/columns/read/1343/was-mcdonaldland-plagiarized-from-the-old-h-r-pufnstuf-kids-tv-show; *Sid & Marty Krofft Television Productions, Inc., et al. v. McDonald's Corporation and Needham, Harper & Steers Inc.*, U.S. Court of Appeals Ninth Circuit, October 12, 1977.

Pg. 92 *"Favorites included: the"*: Not so Secret Menu at the In-N-Out website, http://www.in-n-out.com/secretmenu.asp.

Pg. 93 *"In the mid-1960s, McDonald's traced the origins of the hamburger"*: John Love, *McDonald's: Behind the Arches*, rev. ed. (New York: Bantam Books, 1995), 208.

Pg. 93 *"In 1930, White Castle founders Billy Ingram and Walter Anderson"*: White Castle corporate history, http://www.whitecastle.com.

Pg. 94 *"accomplished only with the dedicated enthusiasm"*: Aileen Pinheiro, comp., "In-N-Out, Inc.: Esther L. Snyder," *The Heritage of Baldwin Park*, vol. 1 (Dallas, Tex.: Taylor Publishing Co., 1981), 242.

CHAPTER 8

Pg. 96 *"In 1973, more than 245 franchise companies"*: Robert L. Emerson, *The New Economics of Fast Food* (New York: Van Nostrand Reinhold, 1990), 23, 62.

Pg. 96 *"had launched more than 32,000 fast-food establishments"*: John A. Jakle and Keith A. Sculle, *Fast Food Roadside Restaurants in the Automobile Age* (Baltimore: Johns Hopkins University Press, 1999), 85.

Pg. 96 *"that sold $9.68 billion"*: National Restaurant Association.

Pg. 96 *"You can find your way across this country using burger joints"*: Charles Kuralt, CBS Morning, as cited in John T. Edge, *Hamburgers & Fries, An American Story* (New York: G.P. Putnam's Sons, 2005).

Pg. 96 *"the fast-food industry was growing"*: This figure averaged from the percentage chain of sales growth reported by the National Restaurant Association between the years 1970 and 1979 and cited in Robert L. Emerson, *The New Economics of Fast Food* (New York: Van Nostrand Reinhold, 1990), 23.

Pg. 97 *"Between 1965 and 1971, Kentucky Fried Chicken"*: Robert L. Emerson, *The New Economics of Fast Food* (New York: Van Nostrand Reinhold, 1990), 7.

Pg. 97 *"Between 1965 and 1973, the number of McDonald's outlets"*: Ibid.

Pg. 97 *"Between 1945 and 1960, the Valley's population"*: Kevin Roderick, "America's Suburb Timeline," The Valley Observed, http://www.Americassuburb.com/timeline.html.

Pg. 97 *"that earned it the nickname"*: Kevin Roderick, *The San Fernando Valley: America's Suburb* (Los Angeles: Los Angeles Times Books, 2001).

Pg. 97 *"North Hollywood drive-through was close to the Hollywood Freeway,"*: Los Angeles City Department of Transportation, Cahuenga Parkway report, http://www.lacity.org/ladot/TopicsAndTales/Freeway4.pdf.

Pg. 98 *"Over 12 Billion Sold."*: "The Burger that Conquered America," *Time*, September 14, 1973.

Pg. 99 *"the chain launched its $50 million 'You Deserve a Break Today' campaign"*: Ibid.

Pg. 99 *"In 1965, McDonald's went public"*: McDonald's corporate history; Ray Kroc with Robert Anderson, *Grinding It Out: The Making of McDonald's* (Chicago: H. Regnery, 1977), 149.

Pg. 99 *"Over the next thirty-five years, the stock split twelve times."* "McDonald's the Rise and Stall," *Businessweek*, March 2, 2003, http://www.businessweek.com/magazine/content/03_09/b3822087_mz017.htm.

Pg. 99 *"In 1969, the year of Taco Bell's IPO,"*: Taco Bell corporate history, http://www.tacobell.com/.

Pg. 99 *"KFC (as it later became known) was listed"*: KFC corporate history, http://www.kfc.com/about/history.asp.

Pg. 99 *"When Carl Karcher took his company public in 1981,"*: Carl Karcher Enterprises corporate history, http://www.ckr.com/about_history.html#80s; Louise Kramer, "Carl Karcher," *Nation's Restaurant News*, February 1996.

Pg. 99 *"In 1967, the Pillsbury Company"*: Eric Berg, "Burger King's Angry Franchisees," *New York Times*, November 14, 1988.

Pg. 99 *"Ralston-Purina, best known as a maker of breakfast cereals and pet foods,"*: "Goodall Forges Buyout," *Nation's Restaurant News,* May 13, 1985.

Pg. 99 *"General Foods Corporation purchased Burger Chef"*: Connie J. Zeigler, "Burger Chef," *Encyclopedia of Indianapolis* (Bloomington: Indiana University Press, 1994), 364.

Pg. 99 *"A successful Indianapolis-based chain"*: Ibid.; John A. Jakle and Keith A. Sculle, *Fast Food Roadside Restaurants in the Automobile Age* (Baltimore: Johns Hopkins University Press, 1999), 120.

Pg. 99 *"By 1970, Kentucky Fried Chicken had made"*: "Franchising: Too Much, Too Soon," *Businessweek,* June 27, 1970.

Pg. 99 *"McDonald's announced it was going to increase the price"*: Michael Karl Witzel, *The American Drive-In Restaurant* (St. Paul: MBI Publishing, 1994), 175.

Pg. 99n *"Burger King waited until May 17, 2006,"*: "BK Sets IPO Price," *Nation's Restaurant News,* May 15, 2006.

Pg. 101 *"Two years after General Foods purchased Burger Chef,"*: Remembering Burger Chef, http://www.waymarking.com/cat/details.aspx?f=1&guid=65766bda–9049–4276–9ee6–1fe3ed9e6d1f&exp=True.

Pg. 101 *"In 1972, General Foods took a $75 million loss"*: John Love, *McDonald's: Behind the Arches,* rev. ed. (New York: Bantam Books, 1995), 279.

Pg. 101 *"and a decade later sold the chain"*: "Hardee's to Buy Burger Chef," *New York Times,* December 10, 1981.

Pg. 101 *"McDonald's actually picked up the pace"*: Robert L. Emerson, *The New Economics of Fast Food* (New York: Van Nostrand Reinhold, 1990), 10.

Pgs. 101–102 *"As Robert McKay, the general manager of Taco Bell,"*: "A Promising Mañana," *Forbes,* August 1, 1977.

Pg. 102 *"beef prices were rising sharply."*: John Mariani, *America Eats Out* (New York: William Morrow & Co., 1991), 174.

Pg. 102 *"Over the next few years, nearly all of them began"*: Ibid., 174, 176.

Pg. 102 *"McDonald's added its Quarter Pounder;"*: McDonald's history listing, http://www.mcdepk.com/50/downloads/history_listing.pdf.

Pg. 102 *"Wendy's began offering stuffed baked potatoes,"*: John Mariani, *America Eats Out* (New York: William Morrow & Co., 1991), 176.

Pg. 102 *"Burger King launched its 'Have It Your Way' campaign"*: "Jack v. Mac," *Time,* May 5, 1975; Burger King corporate history of advertising, http://www.bk.com/companyinfo/corporation/history.aspx.

Pg. 102 *"Watch out McDonald's!"*: "Jack v. Mac," *Time,* May 5, 1975.

Pg. 103 *"In some areas, local ordinances forbade fast-food restaurants"*: John A. Jakle and Keith A. Sculle, *Fast Food Roadside Restaurants in the Automobile Age* (Baltimore: Johns Hopkins University Press, 1999), 160.

Pg. 103 *"residents of Woods Hole, Massachusetts,"*: Ibid., 159.

Pg. 103 *"In 1975, three community groups protested the efforts"*: "The Fast-Food Furor," *Time,* April 21, 1975.

Pg. 103 *"This is what our country is all about"*: "The Burger That Conquered the Country," *Time,* September 14, 1973.

CHAPTER 9

Pg. 107 *"It's okay on goals to dream big,"*: Karen de Witt, "The Executive Life; A White House Dinner: The Thrill of a Lifetime," *New York Times*, June 21, 1992.

Pg. 111 *"in 1976, when McDonald's posted $3 billion"*: McDonald's history listing, http://www.mcdepk.com/50/downloads/history_listing.pdf.

Pg. 111 *"He died on December 14, 1976;"*: Obituary, *San Gabriel Valley Tribune*, December 15, 1976.

Pg. 111 *"Hubert Eaton, the founder of Forest Lawn,"*: A history of Eaton and Forest Lawn can be found on the DVD *Forest Lawn: The First Hundred Years*, released in 2006 and available at the Forest Lawn–Glendale museum. Information can also be found in Cecilia Rasmussen, "Cemetery Was Not an End, but a Beginning," *Los Angeles Times*, April 16, 2000; Carey McWilliams, *Southern California Country, An Island in the Land* (New York: Duell Sloan & Pearce, 1946), 230–231.

Pg. 112 *"unlike other cemeteries as sunshine is unlike darkness,"*: Eaton composed this statement for his "The Builder's Creed," written in 1917 and cited in Tom Sitton and William Francis Deverell, eds., *Metropolis in the Making: Los Angeles in the 1920s* (Berkeley: University of California Press, 2001), 346.

Pg. 112 *"Forest Lawn's Wee Kirk O' the Heather Church"*: Claudia Luther, "Jane Wyman, 90, Oscar Winner First Wife of Reagan," *Los Angeles Times*, September 11, 2007.

CHAPTER 10

Pg. 116 *"In 1976, the United States was celebrating its bicentennial"*: Robert L. Emerson, *The New Economics of Fast Food* (New York: Van Nostrand Reinhold, 1990), 23.

Pg. 116 *"It's hard enough to sell burgers, fries, and drinks, right,"*: Ellen Paris, "Where Bob Hope Buys His Burgers," *Forbes*, June 24, 1989.

Pg. 118 *"First established in 1958, Sizzler was a pioneer"*: Sizzler corporate history, http://www.sizzler.com/about/our_history.asp.

Pg. 119 *"At about 7:00 p.m. on August 16, 1978, a fire broke out"*: Randy Woods, "Spectacular Fire Blankets Valley in Dense Smoke," *San Gabriel Valley Tribune*, August 17, 1978.

Pgs. 119–120 *"the San Gabriel Valley Tribune reported"*: Ibid.

Pg. 121 *"It had outgrown its signature format:"*: Richard Martin, "In-N-Out Burgers Pulls Away from Drive-thru Only Focus," *Nation's Restaurant News*, June 19, 1989.

Pg. 121 *"I think double drive-throughs are great;"*: Ibid.

Pg. 123 *"The couple lived in a $600,000 estate"*: Andrew Bluth and Chris Knap, "Life of Fast Food, Cars Came to an Early Halt," *Orange County Register*, February 6, 2000.

CHAPTER 11

Pg. 124 *"In January 1979, Rich hired a local architect to work with him"*: Detailed description of the building of the new headquarters from Baldwin Park Historical Society archives.

Pg. 125 *"The national news was focused on the new president's supply-side economic policy"*: Irving Kristol, "The Truth about Reaganomics," *Wall Street Journal*, November 20, 1981; Bernard Gwertzman, "There May Be More to Foreign Policy than Stopping the Soviet," *New York Times*, April 26, 1981.

Pg. 125 *"Their wedding attracted 750 million television viewers"*: Jay Cocks, "Magic in the Daylight," *Time*, April 20, 1981.

Pg. 125 *"Baldwin Park's chamber of commerce began actively recruiting new businesses"*: Kenneth J. Fanucchi, "Renewal Purses, Chief Quits," *Los Angeles Times*, November 26, 1981.

Pgs. 125–126 Description of the newly built headquarters from Baldwin Park Historical Society archives.

Pg. 127 *"Inside, specially selected cow and steer chucks arrived"*: Nancy Luna, "A Burger's Journey," *Orange County Register*, March 31, 2006.

Pg. 127 *"The chain proudly proclaimed that it paid"*: In-N-Out Burger corporate website, http://www.in-n-out.com/statement.asp.

Pg. 127 *"After In-N-Out's inspection, a team of skilled butchers boned and removed the meat."*: Edmund Newton, "Faithful Customers Have No Beef with In-N-Out Burger," *Los Angeles Times*, September 16, 1990; Nancy Luna, "A Burger's Journey," *Orange County Register*, March 31, 2006.

Pg. 127 *"the family-owned chain was selling more than 14 million burgers each year."*: Donald McAuliffe, "Family Affair," *San Gabriel Valley Tribune*, May 12, 1984.

Pg. 128 *"Roughly two dozen local business executives"*: "In-N-Out to Build New Warehouse," *San Gabriel Valley Tribune*, February 11, 1986.

Pg. 128 *"In-N-Out was generating roughly $60 million in sales annually."*: Richard Martin, "In-N-Out Burgers Pulls Away from Drive-thru Only Focus," *Nation's Restaurant News*, June 19, 1989.

Pg. 128 *"put the figure closer to $73 million."*: Estimate provided by Technomic Inc.

Pg. 128 *"The aims set forth by Harry Snyder since the founding of the company"*: Aileen Pinheiro, comp., "In-N-Out Burger, Inc.: Esther L. Snyder," *The Heritage of Baldwin Park*, vol. 1 (Dallas, Tex.: Taylor Publishing Co., 1981), 242.

Pgs. 128–129 *"Every investment banker in the country would love to take them public."*: Deborah Silver, "Burger Worship," *Restaurant & Institutions*, November 1, 1999.

Pg. 129 *"As early as 1986, Rich remarked that he had to deny the IPO rumor"*: "In-N-Out to Build New Warehouse," *San Gabriel Valley Tribune*, February 11, 1986.

Pg. 129 *"In-N-Out is a great vehicle to do something like that,"*: Mark Sachs, "In-N-Out: A Short Menu Means Steady Growth," *San Gabriel Valley Tribune*, August 24, 1992.

Pg. 129 *"On March 16, 1982, approximately three months after moving in,"*: *The Heritage of Baldwin Park*, April 6, 1992, 2–4.

Pg. 129 *"Come join us for an evening of fun,"*: In-N-Out Burger and the Baldwin Park chamber of commerce invitation, September 27, 1990.

Pg. 131 *"Rich tapped longtime friend Richard Boyd,"*: In-N-Out v. Richard Boyd, Richard Boyd v. In-N-Out, BC345657, filed January 10, 2009, 9–10.

CHAPTER 12

Pg. 133 *"In 1961, McDonald's had founded its own Hamburger University"*: Ray Kroc with Robert Anderson, *Grinding It Out: The Making of McDonald's* (Chicago: H. Regnery, 1977), 126.

Pg. 134 *"In 1983, the company moved from the Elk Grove Village"*: John Love, *McDonald's: Behind the Arches*, rev. ed. (New York: Bantam Books, 1995), 149.

Pg. 134 *"The year that In-N-Out University was established, there were about thirty stores"*: Donald McAuliffe, "Family Affair," *San Gabriel Valley Tribune*, May 12, 1984.

Pg. 134 *"In contrast, during the first quarter of 1984 alone, McDonald's had opened"*: "Record Profits at McDonald's," Associated Press, April 24, 1994.

Pg. 134 *"generating $8 billion annually."*: Eric Pace, "Ray A. Kroc dies at 81; Built McDonald's Chain," *New York Times*, January 15, 1984.

Pg. 134 *"In-N-Out Burger was meeting the industry giant's revenues on a store-to-store basis."*: Stacy Perman, "Fat Burger," *Los Angeles*, February 2004.

Pg. 134 *"the average In-N-Out location came 'pretty close' to the volume"*: "In-N-Out to Build New Warehouse," *San Gabriel Valley Tribune*, February 11, 1986.

Pg. 137 *"Before long, the company launched its own newsletter"*: "A Legacy of Great Service," *Grapevine Gazette* (Spring 2003), http://www.betternewsletter.com/newsletter/spring_2003.html.

Pg. 137 *"In-N-Out could save a 'ton of money' "*: Donald McAuliffe, "Family Affair," *San Gabriel Valley Tribune*, May 12, 1984.

Pg. 138 *"establishing an expansive set of benefits"*: In-N-Out Burger corporate website, http://www.in-n-out.com.

Pg. 138 *"the state of California raised its minimum wage"*: California Department of Industrial Relations, "History of California Minimum Wage," http://www.dir.ca.gov/Iwc/MinimumWageHistory.htm.

Pg. 138 *"the Orange County Register called Rich Snyder"*: Andre Mouchard, "Common Sense," *Orange County Register*, December 17, 1993.

Pg. 138 *"Rich had already boosted In-N-Out's starting wages"*: Ibid.

Pg. 138 *"If you lose your workers, you lose your customers,"*: Ibid.

Pg. 138 *"Famously, around the time of President Nixon's 1972 reelection campaign,"*: John Love, *McDonald's: Behind the Arches*, rev. ed. (New York: Bantam Books, 1995), 357.

Pg. 138 *"The fast-food chain's detractors were further angered"*: Ibid.

Pg. 138 *"February 2008, when starting pay for all new In-N-Out associates"*: In-N-Out corporate website, http://www.in-n-out.com/employment_restaurant.asp.

Pg. 139 *"Two years earlier, the chain raised its own minimum wage"*: *Morning Edition*, NPR, October 18, 2006.

Pg. 139 *"At the time, the minimum wage in the state of California was $6.50"*: California Department of Industrial Relations, "History of California Minimum Wage," http://www.dir.ca.gov/Iwc/MinimumWageHistory.htm.

Pg. 139 *"By contrast Wal-Mart, a company with $375 billion in sales"*: Wal-Mart sales for fiscal year ending January 31, 2008, from Wal-Mart, *2008 Annual Report & Proxy to Shareholders*, April 22, 2008.

Pg. 139 *"some ten times greater than In-N-Out's annual revenue"*: In-N-Out's 2007 sales, $395 million, estimated by Technomic Inc.

Pg. 139 *"was paying its full-time hourly workers $10.51,"*: Jeffrey Goldberg, "Selling Wal-Mart," *New Yorker*, April 2, 2007.

Pg. 139 *"By 1989, top store managers earned about $63,000"*: Ellen Paris, "Where Bob Hope Buys His Burgers," *Forbes*, June 24, 1989.

Pg. 139 *"We're blessed with good employees,"*: Richard Martin, "In-N-Out Pulls Away," *Nation's Restaurant News*, June 19 1989.

Pg. 139 *"Some twenty years later, store managers were pulling in"*: In-N-Out Burger corporate website, http://www.in-n-out.com/employment_mgmt.asp.

Pg. 140 *"roughly 75 percent of employees staying on beyond six months."*: These figures are repeated in numerous places, including Robert W. Van Giezen, "Occupational Wages in the Fast-food Restaurant Industry," *Monthly Labor Review* (August 1994).

Pg. 140 *"In the case of In-N-Out Burger, its managers maintained"*: In-N-Out corporate website, http://www.in-n-out.com/employment_mgmt.asp.

Pg. 140 *"We try and maintain the highest quality level possible,"*: Mark Sachs, "In-N-Out: A Short Menu Means Steady Growth," *San Gabriel Valley Tribune*, August 24, 1992.

Pg. 140 *"I believe in you. You are the best."*: Rich Snyder Dedication DVD, Hillview Acres Children's Home, Rich Snyder Cottage, October 30, 2007.

Pg. 140 *"Times are tough,"*: Ibid.

Pg. 143 *"We want them to share some of their insights,"*: John Brinsely, "Motivation and Money," *Los Angeles Business Journal*, July 3, 2000.

CHAPTER 13

Pg. 145 *"In 1989,* Forbes *magazine featured In-N-Out"*: Ellen Paris, "Where Bob Hope Buys His Burgers," *Forbes*, June 24, 1989.

Pg. 146 *"a group of eight San Francisco friends famously ordered"*: The story of blogger What's Up Willy and the famous 100 x 100 was first posted on his blog and then picked up widely elsewhere, http://whatupwilly.blogspot.com/2006/01/in-n-out–100x100.html.

Pg. 147 *"gobbled down Double-Double burgers at the In-N-Out drive-through"*: Nancy Verde Barr, *Backstage with Julia: My Years with Julia Child* (Hoboken, N.J.: Wiley, 2007), 274.

Pg. 148 *"minutes later I drove back 'round and got the same thing again"*: Jill Scott, "Gordon Ramsay Admits Secret Passion for Fast Food Burgers," *Sunday Mail*, April 20, 2008.

Pg. 148 *"uncopyable advantage"*: Paul Westra, a restaurant industry researcher for SG Cowen, quoted in Mike Steere, "A Timeless Recipe for Success," *Business 2.0*, September 1, 2003.

Pg. 150 *"The burger Goliaths doing business in Southern California"*: Richard Martin, "In-N-Out's Size No Measure of Its Stature," *Nation's Restaurant News*, May 7, 1984.

Pgs. 150–151 *"Reported estimates on In-N-Out's advertising budget ranged"*: Rebecca Flass, "T&O Group Triumphs in Review for In-N-Out Burger," *ADWEEK*, July 31, 2001.

Pg. 152 *"After appearing in over eight hundred commercials"*: Douglas Martin, "Dave Thomas, 69, Wendy's Founder, Dies," *New York Times*, January 9, 2002.

Pg. 152 *"the magazine reported that In-N-Out sold fifty-two thousand burgers per month."*: Ellen Paris, "Where Bob Hope Buys His Burgers," *Forbes*, June 24, 1989.

Pg. 152 *"A year earlier, in 1988, he had told the* San Gabriel Valley Tribune,"*: Brian H. Greene, "In-N-Out Eyes Opening Up in San Diego," *San Gabriel Valley Tribune*, October 22, 1988.

Pg. 153 *"We aren't striving to become a household name."*: Deborah Silver, "Burger Worship," *Restaurants & Institutions*, November 1, 1999.

Pg. 153 *"If you have to tell somebody you're something, you're probably not."*: Mike Steere, "A Timeless Recipe for Success," *Business 2.0*, September 2003.

CHAPTER 14

Pg. 155 *"You believe in God and you still enjoyed science?"*: Snyder family home movie; Esther Snyder, interview by Rich Snyder, circa early 1970s.

Pg. 155 *"It began in 1965 as a congregation of about twenty-five parishioners"*: "Brief History of the Calvary Chapel of Costa Mesa," http://www.calvarychapel. com/?page=about; Chuck Smith, "The Complete History of Calvary Chapel," *Last Times*, (Fall 1998), http://www.calvarychapel.com/assets/pdf/LastTimes-Fall1981.pdf.

Pg. 155 *"Smith was a leading figure in the grassroots 'Jesus Movement.'"*: David Di Sabatino, *Jesus People Movement* (Westport, Conn.: Greenwood Press, 1999).

Pg. 155 *"In time, the Calvary Chapel would boast over one thousand congregations"*: Calvary Chapel of Costa Mesa, http://www.calvarychapel.com/?page=about.

Pg. 157 *"In 1984, they established the Child Abuse Fund"*: In-N-Out Burger corporate website, "In Loving Memory of Esther Snyder," http://www.in-n-out.com/esther/; In-N-Out Burger Foundation, http://www.in-n-out.com/foundation.asp.

Pg. 157 *"Each April, canisters were placed in all of the chain's stores"*: "Burger Promotion Aids Abused Children," *San Gabriel Valley Tribune*, June 6, 1986; In-N-Out Burger Foundation, http://shop.in-n-out.com/innout/dept.asp?dept_id=110.

Pg. 157 *"on June 6, 1986, a photograph in the* San Gabriel Valley Tribune*"*: "Burger Promotion Aids Abused Children," *San Gabriel Valley Tribune*, June 6, 1986.

Pg. 158 *"Over the years, the fund distributed millions of dollars."*: In-N-Out Burger corporate website, "In Loving Memory of Esther Snyder," http://www.in-n-out.com/esther/.

Pg. 159 *"Hamburgers are so popular,"*: Stacy Perman, "Fat Burgers," *Los Angeles*, February 2004.

Pg. 159 *"It gets the Christian community pretty excited"*: Tracy Weber, "Religion Is the Meat of Burger Chain's Ad," *Orange County Register*, December 24, 1991.

Pgs. 159–160 *"not everybody that listens to you is a Christian,"*: Ibid.

Pg. 160 *"It would be a real drag to die and be up in front of God"*: Ibid.

Pg. 161 *"In 1980, he backed Republican David Dreier"*: http://newsmeat.com, aggregation of federal filing political contributions.

Pg. 161 *"Soon Rich began contributing tens of thousands of dollars"*: http://newsmeat. com, aggregation of federal filing political contributions.

Pg. 162 *"When he was a child, she told him to always smile,"*: Comments made by Esther Snyder during her eulogy at memorial for Rich Snyder, Phil West, and Jack Sims, held at the Calvary Chapel in Costa Mesa, December 23, 1993 (videotape).

CHAPTER 15

Pg. 165 *"He hired a specially outfitted passenger train"*: Brian H. Greene, "In-N-Out Eyes Opening Up in San Diego," *San Gabriel Valley Tribune*, October 22, 1988.

Pg. 166 *"His goal was to open the first San Diego–area In-N-Out"*: Ibid.

Pg. 166 *"San Diego is a totally new market area for us,"*: Ibid.

Pg. 166 *"Perhaps more important, scores of residents"*: Sharon K. Gillenwater, "Burger Fixation," *San Diego*, March 1990.

Pg. 167 *"And the chain's revenue,"*: Sales figures estimated by Technomic Inc.; average growth rate calculated based on average from annual sales figures.

Pg. 167 *"McDonald's had about 8,576 domestic restaurants"*: Store data compiled by Technomic Inc.

Pg. 167 *"Why pay $2 million for a property"*: Ellen Paris, "Where Bob Hope Buys His Burgers," *Forbes*, June 24, 1989.

Pg. 171 *"In-N-Out was classified as an S-Corporation,"*: "Petition of co-trustee Richard Boyd," *In Re the Matter of Esther L. Snyder Trust–1989*, BP095380 (S.C. Calif. 2005), 16.

Pg. 176 *"In fact, by the summer of 1989, the chain was selling twelve thousand T-shirts"*: Ellen Paris, "Where Bob Hope Buys His Burgers," *Forbes*, June 24, 1989.

Pg. 177 *"On January 31, 1989, the Snyders established an irrevocable family trust:"*: *Esther L. Snyder Trust–1989*, January 31, 1989.

Pg. 177 *"A separate trust, the Lynsi L. Snyder Trust, was set up"*: *Lynsi Snyder Trust–1989*, January 31, 1989.

Pg. 177 *"to make provision for my two sons and other lineal descendants,"*: "Declaration of Esther L. Snyder in Support of Petition to Reform Trust Instrument to Conform to Trustor's Intention," *In Re the Matter of Esther L. Snyder Trust–1989*, KP 005531 (S.C. Calif. 2000).

Pg. 177 *"the Esther L. Snyder Trust was made up of 44,147 shares"*: "Petition of co-trustee Richard Boyd," *In Re the Matter of Esther L. Snyder Trust–1989*, BP095380 (S.C. Calif. 2005),11.

Pg. 177 *"The Lynsi Snyder Trust held 4,370 shares"*: "Respondent and Counter-Petitioner Lynsi Martinez's Verified Petition," *In Re the Matter of Lynsi Snyder Trust–1989*, BP 095 640 (S.C. Calif. 2006), paragraph 19, page 4.

Pg. 178 *"For purposes of this instrument, Traci Lynette Taylor and Terri Louise Perkins"*: *Esther L. Snyder Trust–1989*, January 31, 1989, paragraph 14.1.1, pages 29–30.

Pg. 178 *"Rich was to receive 89.82224 percent"*: *Esther L. Snyder Trust–1989*, January 31, 1989, paragraphs 5.1–5.15.3, pages 3–13.

CHAPTER 16

Pg. 179 *"Then, in 1990, PepsiCo purchased a six-year-old, Michigan-based double drive-through hamburger chain"*: Eben Shapiro, "Coke and Pepsi Skirmishing in Restaurant Trade Press," *New York Times*, May 8, 1992.

Pg. 180n *"During the next seven years,"*: "Onetime PepsiCo chain Hot 'N Now sold for $17K," *Nation's Restaurant News*, March 14, 1005.

Pg. 181 *"For those of us who don't always go to the awards show,"*: *Late Show with David Letterman*, CBS, http://www.ktnl.com/latenight/lateshow/wahoo/index/php/20041029.phtml.

Pg. 182 *"in the company of such notables as Kenneth T. Derr,"*: Donnie Radcliffe and Dana Thomas, "Night of the Dancing Bear," *Washington Post*, June 17, 1992.

Pg. 182 *"I love history, I love our country, and it was all there."*: Karen De Witt, "The Executive Life; A White House Dinner: The Thrill of a Lifetime," *New York Times*, June 21 1992.

Pg. 183 *"There we were, dancing with the President and Mrs. Bush,"*: Ibid.

Pg. 183 *"'So that in years to come,' he explained,"*: Ibid.

Pg. 183 *"Following Rich and Christina's wedding,"*: Richard Martin, "Top In-N-Out Burger Execs Killed in Calif. Plane Crash," *Nation's Restaurant News*, January 3, 1994.

Pg. 186 *"Word reached the community through real estate circles"*: Mark Sachs, "In-N-Out: A Short Menu Means Steady Growth," *San Gabriel Valley Tribune*, August 24, 1992.

Pg. 186 *"In time, In-N-Out became one of Baldwin Park's largest employers."*: Aileen Pinheiro, comp., *The Heritage of Baldwin Park*, vol. 2 (Covina, Ca.: Neilson Press, 1981), 81.

Pg. 186 *"following the city council's rejection of a proposal"*: Steve Tamaya, "Neighbor's Beef Dooms Hamburger Place," *San Gabriel Valley Tribune*, August 19, 1990.

Pg. 186 *"Some 119 businesses"*: Ibid.

CHAPTER 17

Pg. 188 *"In-N-Out was opening about ten new stores each year,"*: Averages calculated from Technomic Inc. data.

Pg. 188 *"In-N-Out Burger's procedures that in December, the chain was granted"*: Mark Kendall and Gillen Silsby, "In-N-Out Burger Chief Dies in Crash," *San Gabriel Valley Tribune*, December 17, 1993.

Pg. 189 *"despite the company's growth spurt, he had no interest in competing"*: Mark Sachs, "In-N-Out: A Short Menu Means Steady Growth," *San Gabriel Valley Tribune*, August 24, 1992.

Pg. 189 *"I think it would be too difficult to maintain quality control."*: Ibid.

Pg. 190 *"Yeah it's close,"*: Transcripts of the pilots' conversation with control tower reported by Jeff Brazil, "O.C. Jet Crash Tape Indicates Pilot Knew Risk," *Los Angeles Times*, July 9, 1994.

Pg. 191 *"At Esther's urging, he had gone to work for In-N-Out corporate"*: Comments made by Esther Snyder during her eulogy at memorial for Rich Snyder, Phil West, and Jack Sims, held at the Calvary Chapel in Costa Mesa, December 23, 1993 (videotape).

Pg. 191 *"in 1986 Sims launched a popular but controversial church"*: Jodi Wilgoren, "Jet Crash Victims Eulogized Amid Tears, Smiles," *Los Angeles Times*, December 24, 1993.

Pg. 191 *"Rich considered West such a close friend that five days before the crash, on December 10, he had a trust drawn up"*: Greg Johnson, "4 In-N-Out Burger Execs Sue Over Trust Language," *Los Angeles Times*, July 17, 1996.

Pg. 192 *"[Richard Snyder] was the type of person who did a lot more"*: In-N-Out Burger company statement reported by Richard Martin, "Top In-N-Out Burger Execs Killed in Plane Crash," *Nation's Restaurant News*, January 3, 1994.

Pg. 192 *"every month, Rich sent an In-N-Out cookout trailer to feed the homeless"*: Rich Snyder Dedication DVD, Hillview Acres Children's Home, Rich Snyder Cottage, October 30, 2007.

Pg. 193 *"The ninety-three-store chain was pulling in about $116 million"*: Technomic Inc.

Pg. 194 *"'That man,' he said, holding back tears of his own, 'was a legend in my mind.'"*: Comments made by Don Miller during his eulogy at memorial for Rich Snyder, Phil West, and Jack Sims, held at the Calvary Chapel in Costa Mesa, December 23, 1993 (videotape).

Pg. 194 *"Richard said, 'Mom, I'm so glad you got to go with us today.'"*: Comments made by Esther Snyder during her eulogy at memorial for Rich Snyder, Phil West, and Jack Sims, held at the Calvary Chapel in Costa Mesa, December 23, 1993 (videotape).

Pg. 195 *"Right now, as our hearts are grieving, and we feel empty inside,"*: Comments made by Christina Snyder Wright during her eulogy at memorial for Rich Snyder, Phil West, and Jack Sims, held at the Calvary Chapel in Costa Mesa, December 23, 1993 (videotape).

Pg. 195 *"he alluded to the accident and spoke of 'God's work.'"*: Comments made by Guy Snyder during his eulogy at memorial for Rich Snyder, Phil West, and Jack Sims, held at the Calvary Chapel in Costa Mesa, December 23, 1993 (videotape).

Pg. 196 *"When Richard was killed"*: Myrna Oliver, "Esther Snyder," *Los Angeles Times*, August 6, 2006.

CHAPTER 18

Pg. 200 *"Every year, since she was two years old, he had made a special date"*: Comments made by Esther Snyder during her eulogy at memorial for Rich Snyder, Phil West, and Jack Sims, held at the Calvary Chapel in Costa Mesa, December 23, 1993 (videotape).

Pg. 201 *"Without a legal arrangement in place removing Guy Snyder"*: Esther L. Snyder Trust–1989, pages 3–4 and paragraph 5.6, page 10.

Pg. 201 *"Among the many tragedies and pieces of unfinished business"*: Esther L. Snyder Trust–1989, paragraph 10.1.1, page 19.

Pg. 202 *"When Rich took it on, it was a nice little place with a '50s style,"*: Stacy Perman, "Fat Burgers," *Los Angeles*, February 2004.

Pg. 202 *"My life has changed quite a bit."*: Comments made by Guy Snyder during his eulogy at memorial for Rich Snyder, Phil West, and Jack Sims, held at the Calvary Chapel in Costa Mesa, December 23, 1993 (videotape).

Pg. 204 *"Guy believes we need to have the time to train people properly"*: Greg Johnson, "More than Fare," *Los Angeles Times*, August 15, 1997.

Pg. 205 *"Puzder went on to work for Carl's Jr."*: Jim Keohane, "Fat Profits," *Conde Nast Portfolio*, January 1, 2008; "Karcher Executive Indicted," Associated Press, February 20, 1989; John Emshwiller, "Carl Karcher Head and 15 Others Charged by SEC," *Wall Street Journal*, April 15, 1988; "Settlement in Carl's Jr. Case," Associated Press, July 26, 1989.

Pg. 206 *"the law firm's trust work resulted in 'significant confusion, ambiguity and expense'"*: Greg Johnson, "4 In-N-Out Burger Execs Sue Over Trust Language," *Los Angeles Times*, July 17, 1996.

CHAPTER 19

Pg. 208 *"the chain was generating an estimated $133 million in sales."*: Sales figure estimate from Technomic Inc.

Pg. 209 *"$74.3 billion fast-food industrial complex"*: Charlene C. Price, "Foodservice Sales Reflect the Prosperous, Time-Pressed 1990's," *Food Review*, September–December 2000.

Pg. 209 *"McDonald's had swelled to thirteen thousand stores"*: Technomic Inc.

Pg. 209 *"McDonald's opened new stores in Kuwait and Egypt."*: McDonald's corporate history, http://www.mcdonalds.com/corp/about/mcd_history_pg1/mcd_history_pg5.html.

Pg. 209 *"Burger King, which had opened its first international franchise"*: Excerpt from an S–1 SEC filing, filed by Burger King Holdings Inc. on February 16, 2006.

Pg. 209 *"opening of store number ten thousand"*: "Burger King Corporation Announces the Opening of the Company's 10,000th Restaurant," Burger King Corporation press release, November 6, 1998.

Pg. 209 *"He had taken his Southern California chain"*: Carl Karcher Enterprises corporate history, http://www.ckr.com/about_history.html#80s.

Pg. 209 *"Six years later, of the 561 Carl's Jr. restaurants,"*: Anne Michaud, "New Franchise Strategy Unveiled by Carl Karcher Fast-Food Industry," *Los Angeles Times*, November 17, 1990.

Pg. 209 *"the chain was earning $575 million annually."*: Sales figure from Technomic Inc.

Pg. 209 *"He invested heavily in real estate in Anaheim"*: Eric Malnic, "Carl Karcher, 90; Entrepreneur Turned Hot Dog Stand into a Fast-Food Empire," *Los Angeles Times*, January 12, 2008; Joe Keohane, "Fat Profits," *Conde Nast*, February 2008; Kelly Barron, "A New Burger Combo Deal," *Orange County Register*, April 29, 1997; Mark Schoifet, "Carl Karcher, Founder of Carl's Jr. Chain, Dies at 90," *Bloomberg*, January 12, 2008.

Pg. 210 *"After Taco Bell introduced the concept of value pricing"*: "55-cent Big Macs May Ignite Fast-Food Price War," *San Jose Mercury News*, February 27, 1997; "Taco Bell Restarts Value Wars in Fast Food Industry," PR Newswire, June 28, 1996.

Pg. 210 *"By 1997, the fast-food industry reached $109.5 billion"*: Mark D. Jekanowski, "Causes and Consequences of Fast Food Sales Growth," *Food Review*, January–April 1999; Barnaby J. Feder, "McDonald's Still Finds There's Still Plenty of Room to Grow," *New York Times*, January 9, 1994; "Bigger Portions Being Thrown as Global Fast-Food Fight Heats Up," *Press-Telegram*, March 8, 1996.

Pg. 210 *"in 1996 McDonald's introduced the Arch Deluxe,"*: Arthur Lubow, "Steal This Burger," *New York Times*, April 19, 1998; Stuart Elliott, "Another Agency Creates Ads for McDonald's Adult Burger," *New York Times*, August 16, 1996.

Pg. 211 *"there were 116 In-N-Out Burger drive-throughs"*: Sales figures estimated by Technomic Inc.

Pg. 211 *"Taylor began working at In-N-Out Burger in 1984"*: Ex Parte Application to Compel Richard Boyd to Surrender Trust Property to Mark J. Taylor, BP 0956395 (S.C. Calif.), 12.

Pg. 212 *"police officers in Claremont"*: The People of the State of California v. Defendant Harry Guy Snyder, 6PM00465 (M.C. Pomona 1999).

Pg. 216 *"On January 23, 1997, Guy and Lynda Snyder's divorce became final."*: Los Angeles Supreme Court Civil Court summary.

CHAPTER 20

Pg. 221 *"On October 27, 1997, Tom Wright was stopped at the San Ysidro Port"*: United States of America v. William Thomas Wright, 3:97cr-03353-RBB-1 (U.S. D.C. Calif. 1997).

Pg. 222 *"In December, two months after his arrest, Tom Wright"*: Ibid.

Pg. 222 *"The disease, often cited as the source of King George III"*: T. Cox, N. Jack, S. Lofthouse, J. Watling, J. Haines, M. Warren, "King George III and Porphyria: An Elemental Hypothesis and Investigation," *The Lancet*, vol. 366, issue 9482 (July 23–29, 2005), 332–335.

Pg. 223 *"Guy made arrangements to obtain a legal order"*: Order Approving Trustee's Petition Re Construction of Trust Instrument . . . In Re the Matter of Esther L. Snyder Trust—1989, KP005531 (S.C. Calif. 1997).

Pg. 223 *"Three cotrustees were named:"*: Ibid., 2.

Pg. 223 *"The new agreement was set up in such a way that no new successor"*: Ibid.

Pg. 224 *"Others concluded that it was Lynsi's mother, Lynda,"*: In-N-Out v. *Richard Boyd and Michael Anthony Concrete* and *Richard Boyd v. INO et al.*, BC345657 (S.C. Calif 2006), paragraph 5, page 4.

Pg. 224 *"In February 1997, the chain unseated Wendy's Old Fashioned Hamburgers"*: Daniel Puzo, "17th Annual Choice in Chains," *Restaurants & Institutions*, February 1997.

Pg. 224 *"Wendy's had about 4,757 stores across the United States,"*: Number of stores compiled by Technomic Inc.

Pg. 224 *"was competing with all of the large national chains,"*: Daniel Puzo, "17th Annual Choice in Chains," *Restaurants & Institutions*, February 1997.

Pg. 224 *"At seventy-seven, she was inducted into the California Restaurant Association's Hall of Fame."*: California Restaurant Association.

Pg. 226 *"In the spring of 1999, he came down with pneumonia,"*: Described in the investigator's report, County of Los Angeles Department of Coroner, case #99–087274; information source Lynda Snyder ex-wife of decedent, Deputy Brooks Lancaster Police station report #99–50153–1126–491, and Antelope Valley Hospital, December 5, 1999.

Pg. 227 *"throughout 1999, Lynda was involved in Guy's hospitalizations"*: Ibid.

Pg. 228 *"Sheriff's deputies, who had arrived on the scene"*: Andrew Bluth and Chris Knap, "Life of Fast Food, Cars Came to an Early Halt," *Orange County Register*, February 6, 2000.

Pg. 228 *"He was a hell of a guy."*: Stacy Perman, "Fat Burgers," *Los Angeles*, February 2004.

Pg. 228 *"The Los Angeles County Coroner's office performed an autopsy"*: County of Los Angeles Department of Coroner, case # 99–08274, December 5, 1999.

Pg. 228 *"On February 6, 2000, two months after Guy died, the story of his arrest"*: Andrew Bluth and Chris Knap, "Life of Fast Food, Cars Came to an Early Halt," *Orange County Register*, February 6, 2000.

Pg. 229 *"At the time of Guy Snyder's death, In-N-Out Burger had grown"*: Ibid.

Pg. 229 *"It was earning an estimated $212 million"*: Sales figures estimated by Technomic; growth percentages averaged taken from Technomic sales figures.

CHAPTER 21

Pg. 230 *"Don't let her age fool you,"*: Greg Hernandez, "In-N-Out to Stay Private GM Says," *Los Angeles Times*, December 10, 1999.

Pg. 230 *"Esther is very, very tired."*: Greg Hernandez, "Family-Owned In-N-Out at a Crossroads," *Los Angeles Times*, July 2, 2000.

Pg. 231 *"I'm sure there would be no shortage of potential buyers,"*: Greg Hernandez, "In-N-Out Exec's Death Raises Succession Questions," *Los Angeles Times*, December 9, 1999.

Pg. 231 *"only 30 percent of all family businesses"*: The Family Business Forum at the University of North Carolina.

Pg. 231 *"the $107.1 billion fast-food industry."*: National Restaurant Association.

Pg. 231 *"Johnny Rockets, a successful chain of retro diners."*: Johnny Rockets corporate history; Amy Spector and Richard Martin, "Ronn Teitelbaum, Johnny Rockets Founder, Dies at 61," *Nation's Restaurant News*, September 25, 2000.

Pg. 231 *"But in 1995, with about sixty shops, Teitelbaum sold"*: Amy Spector and

Richard Martin, "Ronn Teitelbaum, Johnny Rockets Founder; Dies at 61," *Nation's Restaurant News*, September 25, 2000.

Pg. 232 *"Twelve years later Red Zone Capital Fund II"*: David Cho, "Snyder Buys Johnny Rockets Diner Chain," *Washington Post*, February 10, 2007.

Pg. 232 *"By then there were 213 stores across the United States"*: "Johnny Rockets Names Lee Sanders New President and CEO," Johnny Rockets press release, May 24, 2007, http://www.johnnyrockets.com/aboutus/press.php?id=160.

Pg. 232 *"news of the 'new economy' propelled by technology."*: The newspaper proclaimed that the Internet was a "gold mine," and headlines trumpeted "Rally Heard Round the World, Dow Jones Industrial Average Skyrockets as Bull Market Continues," and "Strong Job, Pay Figures Fuel Stock Market Rise." All headlines from the *Los Angeles Times*, week of December 4, 1999.

Pg. 232 *"We have a team of people out in our stores that are all similarly committed,"*: Greg Hernandez, "Family-Owned In-N-Out at a Crossroads," *Los Angeles Times*, July 2, 2000.

Pg. 233 *"(Ammerman later went on to serve on the boards of Carl Karcher Enterprises"*: Company information, Carl Karcher Enterprises and Quicksilver.

Pg. 233 *"Lynsi would not begin to receive shares"*: Declaration of the Esther L. Snyder Trust–1989, paragraph 5.9, page 11; Declaration of Lynsi Snyder Trust–1989, paragraph 3.2, page 2.

Pg. 233 *"Lynsi also became the sole beneficiary of the Harry Guy Snyder Testamentary Trust."*: In Re the Estate of Harry Guy Snyder, BP066610 (S.C. Calif. 2003), paragraph 4, page 2.

Pg. 233 *"Its primary assets were a 70 percent interest"*: Ibid., 2–10.

Pg. 234 *"Guy put in a stipulation that his daughter not receive any of his Porsches"*: Will of Harry Guy Snyder, May 18, 1999, 3.

Pg. 234 *"Lynsi was entitled to a third of this trust upon turning thirty,"*: In Re the Estate of Harry Guy Snyder, BP066610 (S.C. Calif. 2003), paragraph B, page 5.

Pg. 234 *"Guy's thirteen-page last will and testament,"*: Will of Harry Guy Snyder, May 18, 1999, 8–13.

Pg. 234 *"Guy did bequeath gifts to a select few."*: Ibid., 2.

Pg. 234 *"Perhaps underlining just how close the two men had become,"*: Ibid., paragraph 5.1, page 9.

Pg. 234 *"Since breaking her hip while in Redding for the store opening"*: In-N-Out v. Richard Boyd and Michael Anthony Concrete and Richard Boyd v. INO et al., BC345657 (S.C. Calif 2006), paragraph 30, page 8.

Pg. 235 *"Years earlier she had undergone heart surgery,"*: "Opposition to Motion to Compel Compliance with Subpoena for Deposition of Esther Snyder," In Re the Matter of Esther L. Snyder Trust–1989; declarations of James P. Larsen, MD, and Kenneth R. Jutzy, MD, BP095380, April 4, 2006; signed letter from Dr. Larsen to Esther's attorney James Morris, February 22, 2006.

Pg. 235 *"Owing to her failing health, in February 2000, Esther bowed out"*: Greg Hernandez, "Family-Owned In-N-Out at a Crossroads," *Los Angeles Times*, July 2, 2000.

Pg. 235 *"She's wonderful. She's so into the people of the company."*: Ibid.

Pg. 238 *"mature and had sufficient experience to successfully manage the company."*: In-N-Out v. Richard Boyd and Michael Anthony Concrete and Richard Boyd v. INO et al., BC345657 (S.C. Calif 2006).

Pg. 239 *"it was Lynda who telephoned Esther to inform her that Guy had died."*: As described in the investigator's report, County of Los Angeles Department of Coroner, case #99–087274; information source Lynda Snyder ex-wife of decedent, Deputy Brooks Lancaster Police station report #99–50153–1126–491, and Antelope Valley Hospital, December 5, 1999.

Pg. 239 *"Lynda was said to have become deeply involved"*: *In-N-Out v. Richard Boyd and Michael Anthony Concrete* and *Richard Boyd v. INO et al.*, BC345657 (S.C. Calif 2006), paragraph 5, page 4.

Pg. 239 *" The church's founder Steven A. Radich,"*: Description comes from the Successful Christian Living Church's website, http://www.scliamc.com/ (last accessed July 2008, site now discontinued).

Pg. 240 *"all assets of trusts—including all of the shares of In-N-Out stock"*: *Declaration of the Lynsi Snyder Trust–1989*, paragraphs 3.3, 3.4, pages 3,4; *In-N-Out v. Richard Boyd and Michael Anthony Concrete* and *Richard Boyd v. INO et al.*, BC345657 (S.C. Calif 2006), paragraph 28, page 8; "Petition of Co-Trustee Richard Boyd," *In Re the Matter of Esther L. Snyder Trust–1989*, BP095380 (S.C. Calif 2005), paragraph 9 (a), (b), (c), (d), pages 8–9.

Pg. 240 *"Soon, the couple were said to have become deeply involved in the Successful Christian Living Church."*: *In-N-Out v. Richard Boyd and Michael Anthony Concrete* and *Richard Boyd v. INO et al.*, BC345657 (S.C. Calif 2006), paragraph 49, page 13.

CHAPTER 22

Pg. 242 *"By 2000, the chain had grown to 142 stores,"*: Sales figures estimated by Technomic Inc.

Pg. 242 *"there were nine hundred applicants for seventy positions."*: Holly Skla, "Raw Deal for Workers on Minimum Wage Anniversary," June 25, 2003, http://www.commondreams.org.

Pg. 243 *"Fast-food sales in the United States, well on their way to approaching $150 billion,"*: National Restaurant Association.

Pg. 243 *"McDonald's, which was at one time considered such an American icon"*: David Grainger, "Can McDonald's Cook Again?" *Fortune*, April 14, 2003.

Pg. 243 *"In 2002, fast food ranked dead last on the University of Michigan's American Customer Satisfaction Index"*: University of Michigan's American Customer Satisfaction Index, http://www.umich.edu/news/index.html?Releases/2002/Feb02/chr021902.

Pg. 243 *"Schlosser's best-selling* Fast Food Nation *was published in 2001."*: Eric Schlosser, *Fast Food Nation: The Dark Side of the All-American Meal* (Boston: Houghton Mifflin Books, 2001).

Pg. 243 *"Three years later, Morgan Spurlock released his documentary"*: *Super Size Me*, released May 7, 2004.

Pg. 243 *"Outbreaks of* E. coli*"*: Numerous reports linked fast food with *E. coli* in the early 1990s, including B. P. Bell, et al., "A Multistate Outbreak of Escherichia coli O157:H7-Associated Bloody Diarrhea and Hemolytic Uremic Syndrome from Hamburgers. The Washington Experience," *Journal of the American Medical Association*, November 2, 1994.

Pg. 243 *"mad cow disease"*: Among the many reports about the impact of mad cow disease on fast food was Matt Kranz, "Mad Cow Socks Fast-Food," *USA Today*, May 20, 2003.

Pg. 243 *"connections to obesity"*: "TV, Lots of Fast Food Triple Obesity Risk," http://www.cnn.com/Health, March 10, 2003.

Pg. 243 *"In January 2003, McDonald's posted its first ever quarterly loss"*: "Cowed to Change," *Economist*, April 8, 2003.

Pg. 244 *"CEO Jim Cantalupo oversaw the company's comeback strategy,"*: Ibid.

Pg. 244 *"In 2002, McDonald's established its 'Dollar Menu.'"*: Melanie Warner, "Salads or No, Cheap Burgers Revive McDonald's," *New York Times*, April 19, 2006.

Pg. 244 *"McDonald's poached four-star chef Dan Coudreaut"*: "Big Mac: Inside the McDonald's Empire," CNBC, July 25, 2007.

Pg. 244 *"In 2004, the company announced a complete store redesign"*: Pallavi Gogoi, et al., "Mickey D's McMakeover," *Businessweek*, May 15, 2005.

Pg. 245 *"In 2003, sales reached an estimated $302 million,"*: Sales estimates from Technomic Inc. and percentage growth rate based on those figures.

Pg. 245 *"In May 2003, the company quietly settled a lawsuit"*: Andrew Galvin, "Irvine, Calif.-Based In-N-Out Burger Settles E. Coli Lawsuit," *Knight-Ridder Tribune*, May 6, 2003.

Pg. 245 "Vegetarians In Paradise, *staged a protest"*: "Say no to In-N-Out Burger," March 1, 2003, http://www.vegparadise.com/news28.html.

Pg. 245 *"their earnest protest was 'greeted with a giant wave of indifference.'"*: "Check Out a Burger at Your Local Library," April 2, 2003, http://www.vegparadise.com/news29.html.

Pg. 245 *"Becoming literate may help our youth learn about healthy eating,"*: Ibid.

Pg. 246 *"The French are jealous"*: John Tierney, "The Big City; French Chefs Cast an Eye on Le Big Mac," *New York Times*, July 18, 2000.

Pg. 246 *"Boulud's entry the 'debut of the gourmet hamburger.'"*: John Kessler, "High-falutin Hamburgers," *Atlanta Journal-Constitution*, April 7, 2005.

Pg. 246 *"The Old Homestead, reputedly the oldest steakhouse in Manhattan,"*: Greg Morago, "Burger Boom," *Hartford Courant*, August 29, 2006.

Pg. 247 *"And the French came to embrace the hamburger too"*: Jane Sigal, "In Paris, Burgers Turn Chic," *New York Times*, July 16, 2008.

Pg. 248 *"a great California institution."*: Patrick McGeehan, "The Red Carpet Leads to Drive-Through," *New York Times*, March 7, 2004.

Pg. 248 *"a 'sprinkling of magic dust'"*: Josh Sens, "Prix Fixe to the People Thomas Keller Goes Populist with His New Restaurant, Ad Hoc," *San Francisco*, January 2007.

Pg. 249 *"Did just that at In-N-Out Burger / No pickles, no onions, no playin,'"*: Andre Nickatina, "Cadillac Girl," *Hell's Kitchen* (Million Dollar Dream, 2002).

Pg. 249 *"The Big Lebowski,"*: directed by Joel and Ethan Coen (Polygram and Working Title Films, 1998).

Pg. 250 *"heading straight to an In-N-Out"*: Patrick McGeehan, "The Red Carpet Leads to Drive-Through," *New York Times*, March 7, 2004.

Pg. 250 *"a sixty-two-year-old Texas businessman named James Van Blaricum"*: David Wethe, "Burger Dreams Toppled in In-N-Out Court Clash," *Dallas Business Journal*, March 22, 2002.

Pgs. 250–251 *"Van Blaricum allegedly set out to reproduce In-N-Out's extraordinary success"*: Ibid.; Steve McLinden, "Irving Man Denies He Copied Chain's Burgers," *Fort Worth Star-Telegram*, April 19, 2002; (U.S. D.C. Calif.), 8:01-cv-00944-DOC-An.

Pg. 251 *"The suit was settled on February 14, 2002."*: In-N-Out Burgers v. Lightning Burgers, et al. (U.S. D.C. Calif. 2000).

Pg. 251 *"couldn't prove anything."*: Steve McLinden, "Irving Man Denies He Copied Chain's Burgers," *Fort Worth Star-Telegram*, April 19, 2002.

Pgs. 251–252 *"Led by the company's longtime general counsel Arnold Wensinger,"*: Stephen Gregory, "Lessons and Insight on Southland Businesses, Outsized; Small Firms Feel Pinch of More Aggressive Trademark Policing," *Los Angeles Times*, March 7, 1998.

Pg. 252 *"In December 2000, In-N-Out Burger filed a federal lawsuit"*: Caleb Correa, "Double Double Talk," *Phoenix New Times*, December 21, 2000; "Lawsuit Claims Burger Infringement," Associated Press, June 1, 2001; *In-N-Out Burger v. Whataburger Inc.* (U.S. D.C. Ariz. 2000), 2:00-cv-02285-JAT.

Pg. 252 *"In-N-Out took its trademarks very seriously."*: Ibid.

Pg. 252 *"In-N-Out sued them in September 2003, days after the Rizzas opened an In & Go Burger"*: Edward Russo, "Burger Chain Says 'In & Go' Is Out of Line," *Register-Guard*, September 12, 2003.

Pg. 253 *"Rather than fight it out in court, the Rizzas settled with In-N-Out."*: Edward Russo, "After In-N-Out Feud, Owner Sticks with & Burger," *Register-Guard*, October 13, 2003.

Pg. 253 *"But shortly after Chadder's opened, In-N-Out began receiving inquiries from customers"*: "Memorandum Decision and Order Granting in Part Plaintiff's Motion for Temporary Restraining Order," *In-N-Out Burger v. Chadders Restaurant . . . and Chad Stubbs*, 2:07-CV–394 TS (C.D. Utah 2007); Sara Israelsen, "Chadder's Eatery in American Fork Denies Imitating In-N-Out Burgers," *Deseret Morning News*, June 26, 2007.

Pg. 253 *"Once there, In-N-Out's attorney stood in line and ordered"*: "Memorandum Decision and Order Granting in Part Plaintiff's Motion for Temporary Restraining Order," *In-N-Out Burger v. Chadders Restaurant . . . and Chad Stubbs*, 2:07-CV–394 TS (C.D. Utah 2007), 3.

Pg. 254 *"News of the lawsuit spread quickly."*: "Memorandum Decision and Order Granting in Part Plaintiff's Motion for Temporary Restraining Order," *In-N-Out Burger v. Chadders Restaurant . . . and Chad Stubbs*, 2:07-CV–394 TS (C.D. Utah 2007); Grace Leong, "A.F. Chadder's Restaurant Sued," *Daily Herald*, June 20, 2007.

Pg. 254 *"We don't want to be known as a bully, especially in the legal field."*: Stephen Gregory, "Lessons and Insight on Southland Businesses, Outsized; Small Firms Feel Pinch of More Aggressive Trademark Policing," *Los Angeles Times*, March 7, 1998.

CHAPTER 23

Pg. 255 *"Among those gathered at the pre-opening party was Harry Snyder's nephew"*: Deirdre Newman, "In-N-Out Reaches Milestone in Temecula: 200th Store," *North County Times*, December 30, 2005.

Pg. 255 *"the chain was generating an estimated $350 million in sales"*: Sales estimates, growth percentage based on sales, and rankings from Technomic Inc.

Pg. 256 *"In-N-Out executive and Snyder family friend Richard Boyd filed a lawsuit"*: *Richard Boyd v. In-N-Out Burger*, BC344043, December 7, 2005.

Pg. 256 *"She owned 23.59 percent of the corporation's stock independent of the trusts"*: "Petition of Co-trustee Richard Boyd," *In Re the Matter of Esther L. Snyder Trust–1989*, BP095380 (S.C. Calif. 2005), 16.

Pg. 256 *"Esther's share,"*: "Lynsi Martinez's cannibus memorandum in reply to Boyd's omnibus opposition to Lynsi Martinez's demurrers," *In Re the Matter of Esther L. Snyder Trust–1989*, BP095640 (S.C. Calif. 2006).

Pg. 256 *"the family trusts, valued at $450 million."*: "Petition of Co-trustee Richard

Boyd," *In Re the Matter of Esther L. Snyder Trust–1989*, BP095380 (S.C. Calif. 2005), paragraph 8, page 8.

Pg. 256 *"Boyd contended that she was colluding with Mark Taylor"*: "Plaintiffs Ex Parte Application for Temporary Restraining Order and Order to Show Cause Re Preliminary Injunction," *Richard Boyd v. In-N-Out*, BC 345506, January 5, 2006, 4.

Pg. 257 *"We're in good shape,"*: John Pomfret, "In California, Internal Lawsuits Served Up at Burger Chain," *Washington Post*, January 30, 2006.

Pg. 257 *"There was a brief period of calm as the two parties and their teams of lawyers began settlement discussions."*: *In-N-Out v. Richard Boyd and Michael Anthony Concrete and Richard Boyd v. INO et al.*, BC345657 (S.C. Calif. 2006), paragraph 74, page 20; Lisa Jennings, "In-N-Out Insiders' Lawsuits Roil Icon's Executive Ranks," *Nation's Restaurant News*, January 23, 2006.

Pg. 257 *"On January 5, 2006, Boyd returned to court and re-filed his original suit."*: *Richard Boyd v. In-N-Out Burger*, BC 345506 (S.C. Calif. 2006).

Pg. 257 *"A week later, In-N-Out countersued,"*: "Plaintiff In-N-Out Burgers' Complaint," *In-N-Out v. Richard Boyd and Michael Anthony Concrete and Richard Boyd v. INO et al.*, BC345657 (S.C. Calif. 2006).

Pg. 257 *"On January 30, In-N-Out fired Boyd from his position"*: *In-N-Out v. Richard Boyd and Michael Anthony Concrete and Richard Boyd v. INO et al.*, BC345657 (S.C. Calif. 2006), paragraph 15, page 6.

Pg. 258 *"such bitter infighting and legal wrangling had left the fortunes of the Haft family of Washington D.C.'s estimated $500 million"*: Margaret Webb Pressler, "Herbert Haft Agrees to Settlement," *Washington Post*, April 22, 1997.

Pg. 258 *"the Chicago-based Pritzker family's $15 billion real estate holdings,"*: Jodi Wilgoren, "$900 Million Accord Enables Breakup of Pritzker Dynasty," *New York Times*, January 7, 2005.

Pg. 258 *"and the Mondavi family's $1 billion Napa Valley winery"*: Carol Emert, "Legendary California Wine Company Is Sold," *San Francisco Chronicle*, November 4, 2004.

Pg. 258 *"Boyd's lawsuit, she declared, 'contains outright lies and awful inaccuracies'"*: Lisa Jennings, "In-N-Out Insiders' Lawsuits Roil Icon's Executive Ranks," *Nation's Restaurant News*, January 23, 2006.

Pg. 258 *"The feud began around 2003."*: *Richard Boyd v. In-N-Out Burger*, BC345506 (S.C. Calif. 2006), paragraphs 31, 32, 33, page 7.

Pg. 259 *"According to the filings, Lynsi held weekly prayer meetings"*: *In-N-Out v. Richard Boyd and Michael Anthony Concrete and Richard Boyd v. INO et al.*, BC345657 (S.C. Calif. 2006), paragraph 4, page 3.

Pg. 259 *"Boyd charged that she used Taylor to convey her wishes,"*: *In-N-Out v. Richard Boyd and Michael Anthony Concrete and Richard Boyd v. INO et al.*, BC345657 (S.C. Calif. 2006).

Pg. 259 *"According to court documents, Mark Taylor and Roger Kotch,"*: *Richard Boyd v. In-N-Out Burger* (S.C. Calif. 2006), paragraph 35, page 7.

Pg. 260 *"On April 8, 2004, Boyd asked to meet with Lynsi, Taylor, and a complement of lawyers"*: Ibid., paragraph 36, page 8.

Pg. 260 *"In order to reduce about $47 million in taxes from his estate and pay off the outstanding federal and state taxes owed,"*: "Petition of Co-trustee Richard Boyd," *In Re the Matter of Esther L. Snyder Trust–1989*, BP095380 (S.C. Calif. 2005), pages 13–16; Internal Revenue Service, "Closing Agreement on Final Determination Covering Specific

Matters, Estate of Harry Guy Snyder, Deceased, Richard Boyd and Mark Taylor," August 20, 2005.

Pg. 260 *"However, Boyd claimed that Lynsi refused to attend the meeting."*: *Richard Boyd v. In-N-Out Burger* (S.C. Calif. 2006), paragraphs 37, 38, page 8.

Pg. 260 *"a week later, on April 14, Taylor informed Boyd that Lynsi was asking whether he had reconsidered resigning as cotrustee."*: Ibid., paragraphs 39, 40, 41, 42, 43, 44, pages 8, 9, 10.

Pg. 260 *"From that point on, Boyd contended that he was left out of vice presidents' meetings"*: Ibid., paragraph 46, page 10.

Pg. 260 *"Beginning in June, Boyd was no longer asked to take documents to Esther's home"*: "Declarations of Richard Boyd and Elisa Boyd," *Richard Boyd v. In-N-Out Burger*, BC345506 (S.C. Calif. 2006); "Declaration of Richard Boyd," April 27, 2006; Ibid., paragraph 9, page 3; *In-N-Out v. Richard Boyd and Michael Anthony Concrete* and *Richard Boyd v. INO et al.*, BC345657 (S.C. Calif. 2006), paragraphs 76–103, pages 20–27.

Pg. 261 *"Increasingly, Taylor was exercising his ambitions"*: *In-N-Out v. Richard Boyd and Michael Anthony Concrete* and *Richard Boyd v. INO et al.*, BC345657 (S.C. Calif. 2006).

Pg. 261 *"In a company that had long made decisions by consensus,"*: Ibid.

Pg. 261 *"Sometime during 2004, Taylor reportedly attempted to have Esther put in a home."*: Ibid., paragraph 38, page 10.

Pg. 261 *"The episode did not exactly endear Mark to Esther,"*: Ibid., paragraph 38, page 10.

Pg. 261 *"A campaign of isolating Esther, Boyd charged, began in earnest."*: "Verified Complaint," *Richard Boyd v. In-N-Out Burger*, BC 345 506 (S.C. Calif. 2006), paragraph 25, page 5.

Pg. 261 *"During this time, Boyd insisted that he continued to consult and advise Esther"*: "Declaration of Richard Boyd and Elisa Boyd in Support of Plaintiff's Ex Parte Application for Entry of Temporary Restraining Order and OSC Re Preliminary Injunction," *Richard Boyd v. In-N-Out Burger*, BC345506 (S.C. Calif. 2006), paragraph 9, page 3.

Pg. 262 *"The company initiated an internal investigation"*: *In-N-Out Burgers v. Richard Boyd*, BC345657 (S.C. Calif. 2006), paragraph 13, page 4.

Pg. 262 *"It also maintained that he favored one contractor, Michael Anthony Companies,"*: Ibid., paragraph 11, page 3.

Pg. 262 *"On September 16, 2005, Boyd was given a notice of non-renewal"*: Ibid., paragraph 22, page 6; copy of the letter addressed to Mr. Richard Boyd, Re. Notice of Non-Renewal of Employment Agreement, September 16, 2005.

Pg. 262 *"When Boyd showed Esther the notice, according to court filings, she said that she didn't recognize it."*: "Declaration of Richard Boyd and Elisa Boyd in Support of Plaintiff's Ex Parte Application for Entry of Temporary Restraining Order and OSC Re Preliminary Injunction," *Richard Boyd v. In-N-Out Burger*, BC345506 (S.C. Calif. 2006), paragraph 21, page 5.

Pg. 262 *"Shortly after he left her house, Boyd claimed that Taylor telephoned him,"*: Ibid., paragraph 23, page 5.

Pg. 262 *"Following this episode, Boyd contended that his contact with Esther became less frequent"*: Ibid., paragraphs 28–31, pages 6–7.

Pg. 262 *"After being marginalized for some time, Boyd asserted that by September 2005,*

Esther was essentially a prisoner": In-N-Out v. Richard Boyd and Michael Anthony Concrete and Richard Boyd v. INO et al., BC345657 (S.C. Calif. 2006), paragraph 3, page 3.

Pg. 262 *"Her phone calls and correspondence were screened,"*: "Declaration of Richard Boyd and Elisa Boyd in Support of Plaintiff's Ex Parte Application for Entry of Temporary Restraining Order and OSC Re Preliminary Injunction," BC345506 (S.C. Calif. 2006), paragraphs 28–31, pages 6–7.

Pg. 263 *"On November 5, 2005, In-N-Out Burger retained Grant Thornton,"*: Grant Thornton, special investigative report for In-N-Out Burgers, December 5, 2005.

Pg. 263 *"The firm was given access to Boyd's department offices and files."*: Ibid., introduction, 9, 15, 16.

Pg. 263 *"Five days later, In-N-Out notified Boyd that it would hold a hearing on December 13"*: Mark J. Taylor, "Notice of Meeting of Board of Directors of In-N-Out Burgers, A California Corporation," November 8, 2005.

Pg. 263 *"A month later, Lawrence A. Rosipajla, Grant Thornton's director, delivered a thick, seventy-eight-page report"*: Grant Thornton, special investigative report for In-N-Out Burgers, December 5, 2005.

Pg. 264 *"In January, the board of directors called a special meeting, the purpose of which was to discuss Boyd's termination."*: "Declaration of Robert L. Wallan," BC345506; In-N-Out v. Richard Boyd and Michael Anthony Concrete and Richard Boyd v. INO et al., BC345657 (S.C. Calif. 2006), paragraphs 76–97, pages 20–23.

Pg. 264 *"The battle moved to the chambers of the Superior Court in downtown Los Angeles."*: In-N-Out v. Richard Boyd and Michael Anthony Concrete and Richard Boyd v. INO et al., BC345657 (S.C. Calif. 2006), paragraph 121, page 31; "Trustee of Family Fortune Claims Defamation," Fagelbaum & Heller press release, March 15, 2006.

Pg. 265 This entire section comes from the declaration of Michael Anthony Madrid, signed and dated on October 31, 2005. Note: According to the Declaration of Richard Kemnitzer, he visited Michael Anthony Madrid's office in October 2005 and November 17, 2005. "Declaration of Richard Kemnitzer," Richard Boyd v. In-N-Out Burger, BC345506 (S.C. Calif. 2006).

Pg. 266 *"On March 30, Judge Mitchell Beckloff suspended Boyd as cotrustee of the family trusts,"*: In the Matter of Lynsi L. Martinez Trust–1989, BP 095640, Ruling on Submitted Matter, Superior Court of California, County of Los Angeles, March 30, 2006.

Pg. 266 *"Then, on April 5, the burger chain suffered a setback"*: Fagelbaum & Heller press release; Aurelio Munoz ruling, April 5, 2006.

Pg. 266 *"This is an action for a wrongful discharge,"*: "Motion to Strike Portions of the Cross Complaint," In-N-Out Burgers v. Richard Boyd et al., BC345657 (S.C. Calif. 2006).

Pg. 266 *"Lynsi, Taylor, and In-N-Out filed a motion in February"*: Superior Court of California, County of Los Angeles, April 28, 2006.

Pg. 266 *"Boyd claimed that Lynsi did very little work at the company"*: This description comes variously from the following filings: In the Matter of Harry Guy Snyder Testamentary Trust, BP095639 (S.C. Calif. 2006); In-N-Out v. Richard Boyd and Michael Anthony Concrete and Richard Boyd v. INO et al., BC345657 (S.C. Calif. 2006).

Pg. 267 *"She is often late for work and meetings or does not show up at all."*: Ibid.

Pg. 267 *"Boyd himself reportedly became ensnared in Lynda's wrath"*: In-N-Out v. Richard Boyd and Michael Anthony Concrete and Richard Boyd v. INO et al., BC345657 (S.C. Calif. 2006), paragraph 30 page 8.

Pg. 268 *"Married to Traci Perkins, Taylor considered himself Guy's son-in-law"*: In the

Matter of Harry Guy Snyder Testamentary Trust, BP095639 (S.C. Calif. 2006); *In-N-Out v. Richard Boyd and Michael Anthony Concrete* and *Richard Boyd v. INO et al.,* BC345657 (S.C. Calif. 2006).

Pg. 268 *"The renovation and landscaping were so all-encompassing":* Notes of meeting from Glendora Planning Commission, city staff, City Design Review Committee, City Planning Commission, and City Council, September 2, 2003–September 16, 2003.

Pg. 268 *"Boyd later claimed that Taylor's remodeling project cost $2 million":* In-N-Out v. Richard Boyd and Michael Anthony Concrete* and *Richard Boyd v. INO et al.,* BC345657 (S.C. Calif. 2006), paragraphs 41–44, pages 11–12.

Pg. 269 *"According to Boyd's filings, Taylor was a 'lackluster manager.'":* "Petition of Co-trustee Richard Boyd," *In Re the Matter of Esther L. Snyder Trust–1989,* BP095380 (S.C. Calif. 2005), paragraph 8, page 7.

Pg. 269 *"Boyd insisted that Taylor's ambitions overshadowed his abilities.":* In the Matter of Harry Guy Snyder Testamentary Trust,* BP095639 (S.C. Calif. 2006); *In-N-Out v. Richard Boyd and Michael Anthony Concrete* and *Richard Boyd v INO et al.,* BC345657 (S.C. Calif. 2006).

Pg. 269 *"In his own court filings, Taylor denied Boyd's allegations.":* "Respondent and Counter Petitioner Lynsi Martinez's Verified Petition: (1) to remove Richard Boyd as Trustee; (2) For appointment of Mark Taylor as trustee; (3) in the alternative, for appointment of Northern Trust Bank of California; (4) for an accounting by trustee Richard Boyd, and (5) For attorneys' fees and costs . . .," *In the Matter of Harry Guy Snyder Testamentary Trust,* BP095639 (S.C. Calif. 2006).

CHAPTER 24

Pg. 270 *"At In-N-Out corporate headquarters in Irvine, the inner circle fumed":* In the Matter of Harry Guy Snyder Testamentary Trust,* BP095639 (S.C. Calif. 2006), 7.

Pg. 270 *"In-N-Out Lawsuit Exposes Family Rift,":* Ronald D. White, "In-N-Out Lawsuit Exposes Family Rift," *Los Angeles Times,* January 7, 2006; John Pomfret, "Iconic In-N-Out Battles Executive over Firm's Direction," *Washington Post,* January 30, 2006.

Pg. 271 *"The chain was a 'state treasure,'":* "Double-Double Trouble," *Los Angeles Times,* January 18, 2006.

Pg. 271 *"The whole thing seems sordid, ugly and, worst of all, familiar,":* Joe Christiano, "Wherefore the Double Double," *Los Angeles Times,* February 12, 2006.

Pg. 272 *"The $380 million company":* Sales figure estimated by Technomic Inc.

Pg. 272 *"The Los Angeles–based firm operated":* Sitrick & Company website, http://www.sitrick.com.

Pg. 272 *"charging $695 an hour for what it liked to boast was its expertise in shaping news and public opinion.":* Stevey Oney, "Call Mike Sitrick," *Los Angeles,* July 2006.

Pg. 272 *"the Catholic Archdiocese of Los Angeles during the pedophile scandal":* Carla Hall, "L.A. Archdiocese Enlists Services of Top PR Firm," *Los Angeles Times,* May 30, 2001.

Pg. 272 *"as well as embattled talk show host Rush Limbaugh":* Peter Whoriskey, "Rush Limbaugh Turns Himself In on Fraud Charge in Rx Drug Probe," *Washington Post,* April 29, 2006.

Pg. 272 *"A kind of 'bull in the china shop'"*: Stevey Oney, "Call Mike Sitrick," *Los Angeles*, July 2006.

Pg. 272 *"On March 31, 2006, an article billed as a behind-the-scenes"*: Nancy Luna, "Behind the Scenes at In-N-Out," *Orange County Register*, March 31, 2006.

Pg. 273 *"Esther's attorney bluntly told the* Orange County Register *that Esther would likely side with her granddaughter."* Nancy Luna, "Ailing Leader of In-N-Out Burger Dies," *Orange County Register*, August, 5, 2006.

Pg. 273 *"In a notarized statement dated January 12, 2006, and signed by Esther,"*: "Statement of Esther L. Snyder Concerning Richard C. Boyd," January 12, 2006.

Pg. 274 *"Eleven days later, on January 23, Boyd's attorneys secured a signed declaration from Esther refuting many of In-N-Out's assertions."*: "Declaration of Esther L. Snyder," January 23, 2006.

Pg. 274 *"In February, Boyd moved to have Esther give a deposition."*: "Opposition to Motion to Compel Compliance with Subpoena for Deposition of Esther Snyder," *In Re the Matter of Esther L. Snyder Trust–1989*; declarations of James P. Larsen, MD, and Kenneth R. Jutzy, MD, BP095380, April 4, 2006; signed letter from Dr. Larsen to Esther's attorney James Morris, dated February 22, 2006.

Pg. 275 *"There was no shortage of legal muscle-flexing."*: Ibid.

Pgs. 275–276 *"Esther's niece Alice Meserve Manas signed a court declaration describing Esther as anything but happy"*: "Declaration of Alice Meserve Manas," BP095380 signed April 25, 2006.

Pg. 276 *"in addition to being named to the board"*: "Declaration of Richard Boyd," January 30, 2006.

Pgs. 276–278 *"a transcript of a thirty-four-minute recorded telephone conversation between Esther Snyder and Elisa Boyd,"*: "Declaration of Elisa Boyd, (authenticating Transcript of April 25, 2006 Telephone Conversation with Esther Snyder)," *In Re the Matter of Esther L. Snyder Trust–1989*, BP095380 (S.C. Calif. 2006).

CHAPTER 25

Pg. 281 *"On Tuesday, June 6, Esther's dear friend Margaret Karcher"*: Mary Rourke, "Margaret Karcher," *Los Angeles Times*, June 9, 2008.

Pg. 281 *"On January 11, 2008, a year and half later, Karcher died"*: Carl Karcher Enterprises, "Carl N. Karcher, Founder of Carl's Jr.(R), Dies," January 11, 2008.

Pg. 282 *"also paid tribute to her grandmother."*: In-N-Out Burger corporate website, "In Loving Memory of Esther Snyder," http://www.in-n-out.com/esther/.

Pg. 283 *"God bless Esther Snyder,"*: "Farewell, Esther Snyder, Founder of In-N-Out Burger," *Fast Company*, August 8, 2006, http://blog.fastcompany.com/archives/2006/08/07/farewell_esther_snyder_founder_of_innout_burger.html.

Pg. 283 *"A columnist at the* Orange County Weekly *wrote"*: Steve Lowery, "Diary of a Mad County," *Orange County Weekly*, August 2–8, 2006.

Pg. 285 *"If the chain was worth $300 million to $400 million"*: Pat Maio, "In-N-Out Options Are Many: Steady Growth vs. Private Equity IPO," *Orange County Business Journal*, August 14–20, 2006.

Pg. 285 *"It's a total cult restaurant. People would chase that one."*: David Ellis, "An IPO Wish List For 2007," http://www.cnn.com/money.com, December 6, 2006.

Pg. 285 *"Starbucks chairman Howard Schultz released a memo"*: Starbucks memo, February 14, 2007; Starbucks chairman warns of "the commoditization of the Starbucks experience," first posted February 23, 2007, http://starbucksgossip.com/.

Pg. 286 *"Mark Taylor had assumed the role of president of In-N-Out Burger."*: Nancy Luna, "Ailing Leader of In-N-Out Burger Dies," *Orange County Register*, August 5, 2006.

Pg. 286 *"Taylor was now the sole trustee of the Snyder family trusts"*: In the Matter of *Lynsi L. Martinez Trust–1989*, BP095640 (S.C. Calif. 2006).

Pg. 286 *"In one of Richard Boyd's filings, he had described this scenario"*: "Plaintiffs Ex Parte Application for Temporary Restraining Order and Order to Show Cause Re Preliminary Injunction," *Richard Boyd v. In-N-Out*, BC345506 (2006).

Pg. 286 *"the family is absolutely committed to keeping the company private"*: Nancy Luna, "In-N-Out Leaders Say Burger Chain Won't Change," *Orange County Register*, August 8, 2006.

Pg. 287 *"I was just really hungry, and I wanted to have an In-N-Out burger."*: "Paris Hilton Says DUI Arrest Was Nothing," Associated Press, September 7, 2006.

Pg. 287 *"Neil and I would have cheeseburgers with onions,"*: "Say 'Let's Do Lunch' to Anybody in the World," *San Gabriel Valley Tribune*, November 29, 2006.

Pg. 288 *"The store's manager and divisional manager stood at the entrance welcoming guests."*: Alyson Van Deusen, "Now In," *Spectrum*, April 23, 2008.

Pg. 288 *"If you haven't had an In-N-Out burger"*: Ibid.

SELECT BIBLIOGRAPHY

Bagely, Clarence. *History of Seattle from Earliest Settlement to Present Time*, vol. 2. Chicago, Ill.: The S.J. Clark Publishing Company, 1916.

Belasco, Warren James. *Americans on the Road from Autocamp to Motel, 1910–1945*. Cambridge, Mass.: MIT Press, 1979.

Bialecki, Matthew, et al. *The New Bungalow*. Salt Lake City, Utah: Gibbs Smith, 2001.

Blom, J. C. H., and Emiel Lamberts. *History of the Low Countries*. Oxford: Berghahn Books, 2006.

Bodenhamer, David J., and Robert G. Barrows, eds. *The Encyclopedia of Indianapolis*. Bloomington: Indiana University Press, 1994.

Bond County Historical Society. *Bond County History: A History of Bond County, Illinois*. Greenville, Ill.: Bond County Historical Society, 1979.

Brillat-Savarin, Anthelme. *The Physiology of Taste*. Translated by Fayette Robinson. Philadelphia: Lindsay & Blakiston, 1854. Digitized version from the Elizabeth Robins Pennell Collection available from the Library of Congress.

Burton, Jeffery, et al. *Confinement and Ethnicity: An Overview of World War II Japanese American Relocation Sites*. Seattle: University of Washington Press, 2002.

Chernus, Ira. *Eisenhower's Atoms for Peace*. College Station: Texas A&M University Press, 2002.

Cohen, Lizabeth. *A Consumers' Republic: The Politics of Mass Consumption in Postwar America*. New York: Knopf, 2003.

Davis, Mike. *Ecology of Fear: Los Angeles and the Imagination of Disaster*. New York: Metropolitan Books, 1999.

Di Sabatino, David. *The Jesus People Movement: An Annotated Bibliography and General Resource*. Westport, Conn.: Greenwood Press, 1999.

Edge, John T. *Hamburgers & Fries: An American Story*. New York: G.P. Putnam's Sons, 2005.

Emerson, Robert L. *The New Economics of Fast Food*. New York: Van Nostrand Reinhold, 1990.

Faulkner, William. *Intruder in the Dust*. New York: Vintage Books, 1972.

Foster, Mark S. *A Nation on Wheels: The Automobile Culture in America Since 1945*. Belmont, Ca.: Thomson, Wadsworth, 2003.

Fusao Inada, Lawson, ed. *Only What We Could Carry*. San Francisco: California Historical Society, 2000.

Gabler, Neal. *Walt Disney: The Triumph of the American Imagination*. New York: Knopf, 2006.

Halberstam, David. *The Fifties*. New York: Villard Books, 1993.

Hancock, John. *Village of Panama, 100 Years in the Making*. Self-published, 2006.

Heimann, Jim. *Car Hops and Curb Service: A History of American Drive-In Restaurants 1920–1960*. San Francisco: Chronicle Books, 1996.

Israel, Jonathan Irvine. *The Dutch Republic: Its Rise, Greatness and Fall, 1477–1806*. Gloucestershire, U.K.: Clarendon Press, 1995.

Jackman, Ian. *Eat This: 1,001 Things to Eat Before You Diet*. New York: Harper, 2007.

Jakle, John A., and Keith A. Sculle. *Fast Food: Roadside Restaurants in the Automobile Age*. Baltimore, Md.: Johns Hopkins University Press, 1999.

Kroc, Ray, with Robert Anderson. *Grinding It Out*. Chicago, Ill.: H. Regnery, 1977.

Love, John F. *McDonald's: Behind the Arches*, rev. ed. New York: Bantam Books, 1995.

MacDonald, Norbert. *Distant Neighbors: A Comparative History of Seattle and Vancouver*. Lincoln: University of Nebraska Press, 1987.

McWilliams, Carey. *Southern California Country, an Island in the Land*. New York: Duell, Sloan & Pearce, 1946.

Mariani, John. *America Eats Out: An Illustrated History of Restaurants, Taverns, Coffee Shops, Speakeasies, and Other Establishments that Have Fed Us for 350 Years*. New York: Morrow, 1991.

Moon, Youngme, et al. "In-N-Out Burger," Case Study 9–503–096. *Harvard Business Review*, June 30, 2003.

Newman, Jerry. *My Secret Life on the McJob: Lessons from Behind the Counter Guaranteed to Supersize Any Management Style*. New York: McGraw-Hill, 2007.

Pinheiro, Aileen, comp. *The Heritage of Baldwin Park*, 2 vols. Dallas, Tex.: Taylor Publishing Co., 1981; Covina, Ca.: Nielson Press Inc., 1999.

Roderick, Kevin. *The San Fernando Valley: America's Suburb*. Los Angeles: Los Angeles Times Books, 2001.

Schlosser, Eric. *Fast Food Nation: The Dark Side of the All-American Meal*. Boston: Houghton Mifflin, 2001.

Silk, Gerald. *Automobile and Culture*. New York: Harry N. Abrams, Inc., 1985.

Sitton, Tom, and William Francis Deverell, eds. *Metropolis in the Making: Los Angeles in the 1920s*. Berkeley: University of California Press, 2001.

Stanton, Jeffrey. *Venice California: 'Coney Island of the Pacific.'* Los Angeles: Donahue Publishing, 1993.

Starr, Kevin. *California: A History*. New York: Modern Library, 2005.

Talley-Jones, Kathy, and Letitia Burns O'Connor. *The Road Ahead: The Automobile Club of Southern California 1900–2000*. N.p.: R. R. Donnelley & Sons, 2000.

Tennyson, Jeffrey. *Hamburger Heaven: The Illustrated History of the Hamburger*. New York: Hyperion, 1993.

Trager, James. *The Food Chronology: A Food Lover's Compendium of Events and Anecdotes from Prehistory to the Present*. New York: Henry Holt & Company, 1995.

Tygiel, Jules. *The Great Los Angeles Swindle: Oil, Stocks, and Scandal during the Roaring Twenties*. Berkeley: University of California Press, 1996.

Verde Barr, Nancy. *Backstage with Julia: My Years with Julia Child*. Hoboken, N.J.: John Wiley & Sons, 2007.

Witzel, Michael Karl. *The American Drive-In: History and Folklore of the Drive-In Restaurant in American Car Culture*. St. Paul, Minn.: MBI Publishing, 1994.

INDEX